Pediatric Work Physiology

Medicine and Sport

Vol. 11

Series Editor: E. Jokl, Lexington, Ky.
Assistant Editor: M. Hebbelinck, Brussels

 Published for and on behalf of the Research Committee, International Council of Sport and Physical Education, UNESCO

S. Karger · Basel · München · Paris · London · New York · Sydney

Pediatric Work Physiology

Volume Editors
J. BORMS and M. HEBBELINCK, Brussels

72 figures and 47 tables, 1978

S. Karger · Basel · München · Paris · London · New York · Sydney

Medicine and Sport

Cataloging in Publication
Pediatric work physiology / volume editors, J. Borms and M. Hebbelinck. –
Basel; New York: Karger, 1978. (Medicine and sport; v. 11)
Mainly papers presented at two international symposia.
1. Child Development – congresses 2. Physical Education and Training – in infancy and childhood – congresses 3. Exertion – in infancy and childhood – congresses 4. Adaptation, Physiological – in infancy and childhood – congresses
I. Borms, Jan, ed. II. Hebbelinck, Marcel, ed.
III. Title: International symposium on pediatric work physiology
IV. Title: Symposium on pediatric work physiology V. Series
W1 ME649P v. 11 WE 103.3 P371
ISBN 3-8055-2866-3

Printed in Switzerland by Graphische Betriebe Coop Schweiz, Basel
ISBN 3-8055-2866-3

Contents

Growth and Development

Acknowledgement

The Editors want to acknowledge the invaluable assistance of
Dr. James Day, Lethbridge University, Lethbridge, Alberta

Preface

Research on children and adolescents advances on a broad front. The complexity of human development and the necessity to view for instance growth phenomena in individual subjects over extended periods of time require a diligent effort on the part of the investigators. While the investigator's focus may be on particular problems in given samples, he cannot afford to isolate his own view of human development. Hence, there is a need for him to gain perspective, sustain his resolve and be enriched by the findings and perceptions of colleagues in different disciplines, countries and cultures.

A proposal to initiate a series of informal gatherings in the area of children's physiological work capacity, physical fitness, and growth and development was advanced in multi-disciplinary discussions during the Second International Seminar on Ergometry in Berlin, September 1967. Thus, the idea of forming a 'European Group of Pediatric Work Physiology' with the purpose of holding regular symposia was born.

The inaugural Symposium on Pediatric Work Physiology was held in Dortmund in 1968, under the direction of Professor J. RUTENFRANZ. Since this first and successful meeting, subsequent symposia have become a tradition of this group. Ever since, members of this group have met in Libliče, Czechoslovakia (1969), in Stockholm, Sweden (1970), Wingate, Israel (1972), De Haan, Belgium (1973), Seč, Czechoslovakia (1974), Trois-Rivières, Canada (1975) and Bisham Abbey, England (1976).

All of these symposia contributed to the development of an international research community in pediatric work physiology and have established the essential interdisciplinary support to sustain the participants.

While the primary purpose of these symposia is to provide a scientific forum for determinative discussions, a series of publications has marked the events and provided a record of achievement.

One of the reasons to conceive and develop the present volume was to provide through a number of selected papers a written documentation of the ongoing research in the area of children's and adolescent's physiological work capacity, physical fitness and, growth and development. Seventeen of the nineteen selected papers, comprising the contents of this book, were presented at two International Symposia: the sixth sponsored by the group and organized by Professor M. Maček at Seč, Czechoslovakia in June 1974 and the eighth by Dr. C. T. M. Davies at Bisham, England in September 1976.

The papers, for the largest part, are the work of research teams and represent a multi-variable, interdisciplinary perspective, reflecting that the interest and the understanding of children's and adolescent's responses to physical activity have wider implications than the interest of biological or behavioral scientists only. Amidst the diversity of contributions the organization of the present volume resulted in three chapter headings: Physiological Adaptations to Work; Clinical Pathology; Growth and Development. There are, however, common concerns in both techniques and findings among the three chapters. The overriding concern in all the papers and of the investigators is the need to advance understanding and appreciation of the adaptive processes in children and adolescents. There is the conviction that the more is known about the development and possibilities of children and adolescents the better education, medicine and government can serve the needs of the individual child in a world of increasing technology, of dynamic and social forces and of changing environmental conditions.

Brussels, December 1977. Jan Borms and Marcel Hebbelinck

Medicine Sport, vol. 11, pp. 1–28 (Karger, Basel 1978)

Introduction[1]

ERNST JOKL

During the past decade *Pediatric Work Physiology* has become a scientific discipline of its own. It has not only contributed a great deal of original knowledge but also consolidated information derived from researches conducted earlier.

The South African Study

The first scientific survey of growth of physical efficiency was undertaken forty years ago in South Africa. It has ever since served as model for subsequent studies. The fitness testing program in the United States is based upon it. The grids derived from the analysis of its results (fig. 5–7, pp. 9, 10) are physiological equivalents of the age-related IQ tables presented by BINET and SIMON prior to World War I, subsequently modified by LEWIS TERMAN and HANS JÜRGEN EYSENCK. The grids not only document the development of physical efficiency with increasing age but also allow for differences of children's physiques and functional abilities whose interdependence was demonstrated in detail (fig. 11, p. 13). Every modality of physical power grows within 'channels', provided no extraneous influences such as disease, exercise or diet, cause them to 'cross-over'. Insofar as patterns of growth of physical efficiency of boys and girls differ, they represent a category of secondary sex characteristics of their own (table I–III, pp. 23–25).

[1] Part of project supported by grant No. K 10 LM 00009 awarded by US National Library of Medicine, Bethesda, Md.

PHILIPPE ARIÈS

CENTURIES OF
CHILDHOOD

A Social History of Family Life

Translated from the French by
ROBERT BALDICK

VINTAGE BOOKS
A Division of Random House
NEW YORK

LENGTH OF LIFE

A STUDY OF THE LIFE TABLE

by
LOUIS I. DUBLIN, Ph.D.
SECOND VICE-PRESIDENT AND STATISTICIAN
ALFRED J. LOTKA, D.Sc.
ASSISTANT STATISTICIAN, RETIRED
MORTIMER SPIEGELMAN, F.S.A.
ASSISTANT STATISTICIAN,
METROPOLITAN LIFE INSURANCE COMPANY

REVISED EDITION

THE RONALD PRESS COMPANY • NEW YORK

THE CONQUEST
OF EPIDEMIC DISEASE

A Chapter in the History of Ideas

BY
CHARLES-EDWARD AMORY WINSLOW

PRINCETON, NEW JERSEY
PRINCETON UNIVERSITY PRESS • 1943

ENGLISH CHILDREN

SYLVIA LYND

WITH
12 PLATES IN COLOUR
AND
28 ILLUSTRATIONS IN
BLACK & WHITE

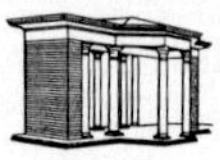

WILLIAM COLLINS OF LONDON
MCMXXXXII

The stratification of test data allows evaluation of efficiency of the best performances among the girls against the weakest performances among the boys (fig. 8–10, pp. 11, 12). Growth of performance efficiency in physical feats demanding speed, strength and endurance occurs in accordance with specific patterns.

Small children were found to possess extraordinarily great powers of endurance. Figure 15, p. 17 shows groups of girls aged 6, 10, 14 and 18, at the end of the 600 yard running test: the 6-year-olds are least, the 18-year-olds most fatigued. Girls attain a physiological maximum of endurance around 14. The South African study was conducted with untrained boys and girls (fig. 12, p. 14).

The 1952 Olympic Games Study

In the interpretation of both growth grids, i.e., those of IQ and of physical efficiency, consideration must be given to the problem of the modifiability of genetic endowment through extraneous influences. The extent to which physical fitness depends not only on training was shown in the Olympic Games Study at Helsinki, Finland in 1952. Using an original statistical approach (fig. 18, p. 19) to the assessment of physical efficiency, the influence upon athletic performances of the Olympic participants of per-capita income, of death rates, of rates of infant mortality, of nutrition and of climate was measured. It became evident that not only in terms of political but also physical power the 'haves' prevail over the 'have-nots' (table IV–VIII, p. 26, 27).

Historical Evidence

The high standards of health and well-being that are at present taken for granted throughout the Western world are unprecedented. Children as physically capable as those participating today in Olympic Games swimming, gymnastics and ice skating competitions did not exist fifty years ago. To place the issue into historical perspective, four examples will be presented: an account of the life of Louis XIV of France (1638–1715); a description of the status of British soldiers around the turn of the 18th century; a reference to child labor in Europe 150 years ago; and the story of the conquest of malnutrition and of the infectious diseases.

Louis XIV

The gorgeous portrait of Louis XIV (1638–1715) by HYACINTHE RIGAUD shows the Grand Monarque in his ermine-lined robes of state, sceptre in hand, arm on hip, gazing out on the world with affable condescension and consummate pose. We see the epitome of majesty: grace, dignity, command (fig. 1). It is a disillusioning experience to turn to the King's private life as described by his doctors. A new and disconcertingly human figure emerges, a man plagued by chronic infirmities of all kinds; pestered by doctors and surgeons; subjected to incredible purges, enemas, and emetics.

The health of even the humblest citizens today is infinitely better protected than that of 'The Sun King' 300 years ago. Throughout his reign Louis XIV was attended by three premier medicins who kept a *Journal de la Santé du Roi*. It describes the illnesses of the royal patient: 'rheumatism, vapours, humours, fistula, insomnia, indigestion, fluxes, headaches, chronic fevers, anthrax, melancholy, urinary difficulties, night sweats, vertigo, erysipelas, colds, colic, bile, acid mouth, chronic toothaches, and, inevitably, a great deal of gout.' It is not difficult to understand why, as PHILIPPE ARIES mentions, 'Louis XIV showed a marked lack of enthusiasm for tennis'.

Fig. 1. Portrait of Louis XIV at the Louvre, Paris by HYACINTHE RIGAUD (1659–1743) 'In the seventeenth century, it was the fortunate man who could not afford medical treatment. The higher one's station in life, the greater became one's medical woes.' (Leon Bernard)

Recruits 1790–1810

A book entitled 'The Enlisting, Discharging and Pensioning of Soldiers During the Years 1790–1819' published in 1839 by HENRY MARSHALL, Deputy Inspector-General of British Army Hospitals, describes that the 100,000 men who during that period enlisted on a long service contract of 21 years were a miserable lot: for a regiment, recruited in 1797 at Cork for embarcation to Buenos Aires, minimum weight was 116 lb. 9 oz. (52 kg), minimum height 5 ft. 5½ in. (1.66 m). Every recruit who complied with these standards was accepted, irrespective of his age. In the French Army minimum weight was 110 lb. (50 kg), minimum height 5 ft. 2 in. (1.575 m). 10% of today's U.S. schoolboys of 12 years of age would thus have qualified for acceptance! Every conceivable disease – epilepsy, palsy, insanity, rheumatism, hernia, blindness, ulcerations – was either simulated or intentionally reproduced in attempts to get out of the army. Nor did lashes deter malingerers from their set purpose. Hernias were produced by incising the scrotum, inserting the stem of a clay pipe and with the assistance of a pal, blowing until the skin was as tight as a drum. Men were quite prepared to lose the sight of one eye if it could get them their discharge. This was done by the application of a caustic, by scraping the surface of the cornea, or by piercing it with a sharp needle and scratching the lense to cause cataract.

Child Labor

At the time of the Industrial Revolution, child-slaves, orphaned and friendless, were supplied throughout Europe in droves to any employer, however brutal, to any employment, however dangerous and degrading (fig. 2). They were deformed and maimed by machinery; they acquired strange new 'industrial' diseases. They were starved and beaten. Children of ten, of seven, of five, even of three, spent long hours of work in the darkness of the mines.

It bespeaks the deep rooted sense of decency of the English that they took the lead in reforming the status of the children so as they had taken the lead in the abolition of slavery. WILLIAM WILBERFORCE formed the Anti-Slavery Society in 1823. ANTHONY ASHLEY COOPER, the 7th Earl of Shaftesbury, Lord of the Admirality under Sir ROBERT PEEL, introduced the 'Ten Hours' Bill' of 1841 which eventually led to the abolition of child labor.

Fig. 2. Children pulling water wagon. Painting by W. G. PJEROV, 1866. Tretyakov Gallery, Moscow. (Bildarchiv Preussischer Kulturbesitz, Berlin)

Fig. 3. Portrait of a mother and 2 children, members of a wealthy family. BENJAMIN DISRAELI, speaking in the House of Commons in 1846 said, 'England is populated by two classes of people; one which produces wealth; the other which consumes it.'

Fig. 4. Title page from *Der Spiegel* with fat girl. Obesity has become a major health problem in the affluent societies.

Rickets

In 1913 Sir John Boyd-Orr assessed the incidence of rickets in Glasgow, Scotland: 40% of the city's school population were found to be afflicted. In a repeat survey in 1936 only 10% of the children in Glasgow had rickets: Vitamin D had been identified by Adolph Windaus who was awarded a Nobel Prize in chemistry in 1928. The Glasgow study was repeated once more in 1964: not a single case of rickets was now found among children who had grown up in Glasgow. But, as a paper by Preece revealed, the incidence of rickets among Asian immigrants to England in 1973 was comparable to what it had been in Glasgow in 1913. Malnutrition is still widespread in Asia and Africa; while obesity has become a major health problem in the affluent societies (fig. 20, p. 21).

Infectious Diseases

In 1900 infectious diseases were responsible for most deaths in the U.S. (fig. 21, p. 21). Fifty years later the chief causes of death were cardiovascular afflictions, malignant tumors, accidents and crimes. During the same period, the lenght of life had increased from 48 to 68 years.

It is against the background of changes such as those referred to above that the favorable status of children with whom the studies described in this book must be assessed. A volume on *Pediatric Work Physiology* could not have been written prior to the turn of our century.

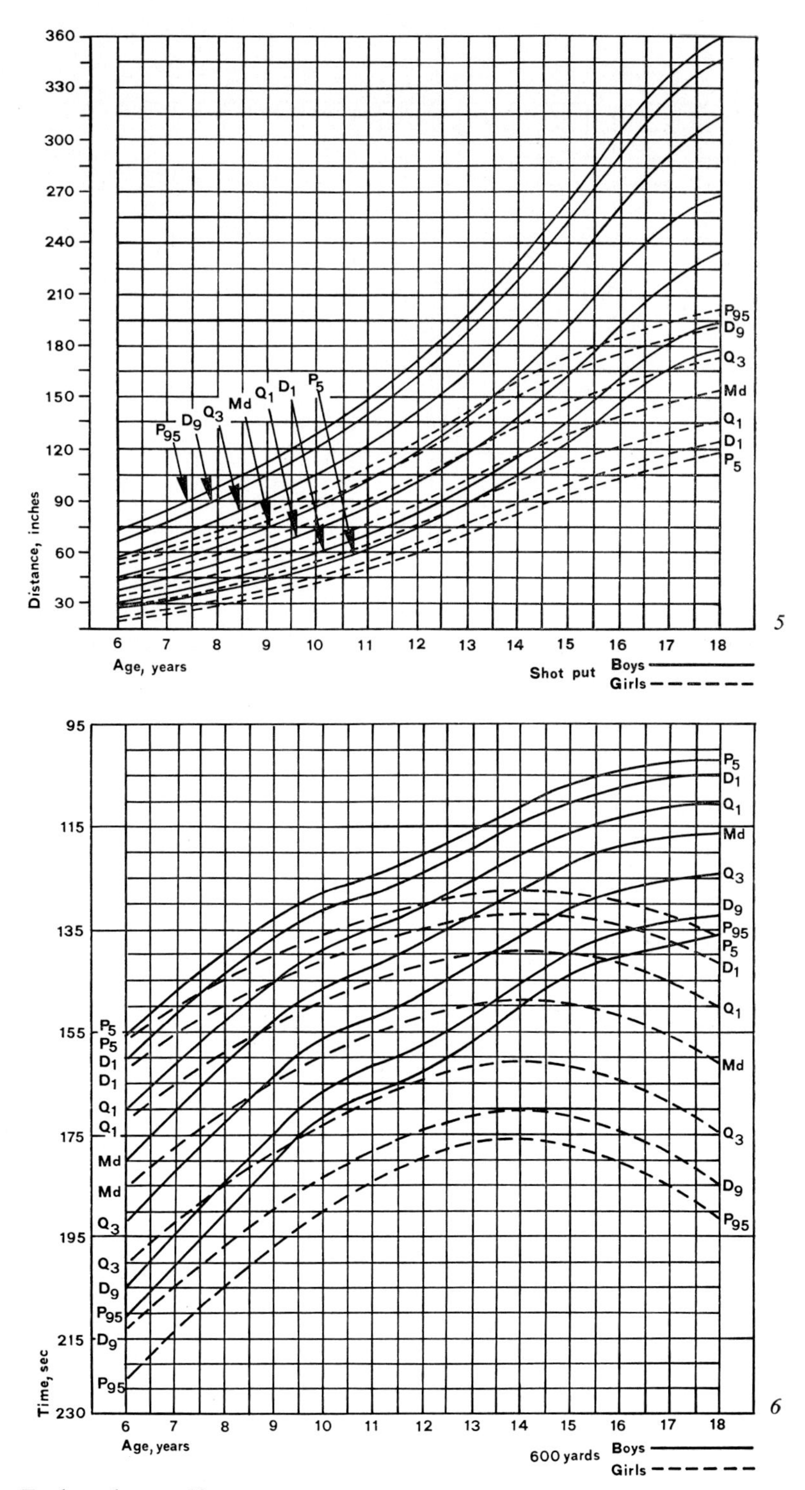

For legends see p. 10

7

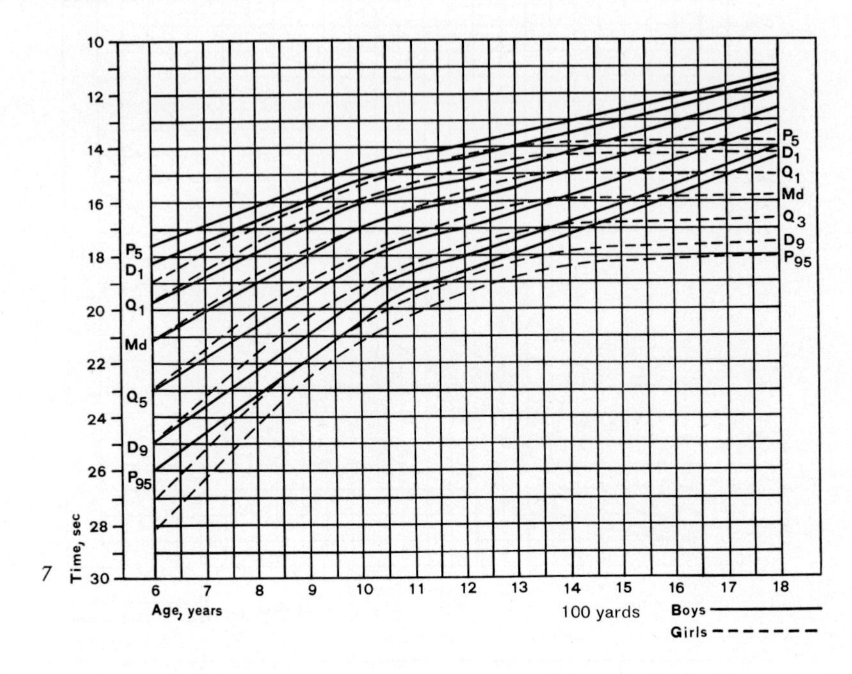

Fig. 5–7. Grids constructed from three tests conducted with 27,654 school children (boys and girls) 6 to 18 years of age: running 100 and 600 yards, and putting the 12 lb. shot. The drawn-out lines are plotted from the boys'; the interrupted lines from the girls' performances. The grids show averages (Md), quartiles (Q_1 and Q_3), deciles (D_1 and D_9) and percentiles (P_5 and P_{95}). The shot put grid shows that performance differences between boys and girls increase with increasing age. In the 100 yards grid differences between boys and girls are small throughout; in the 600 yards grid the direction of the girls' channels changes at age 13.

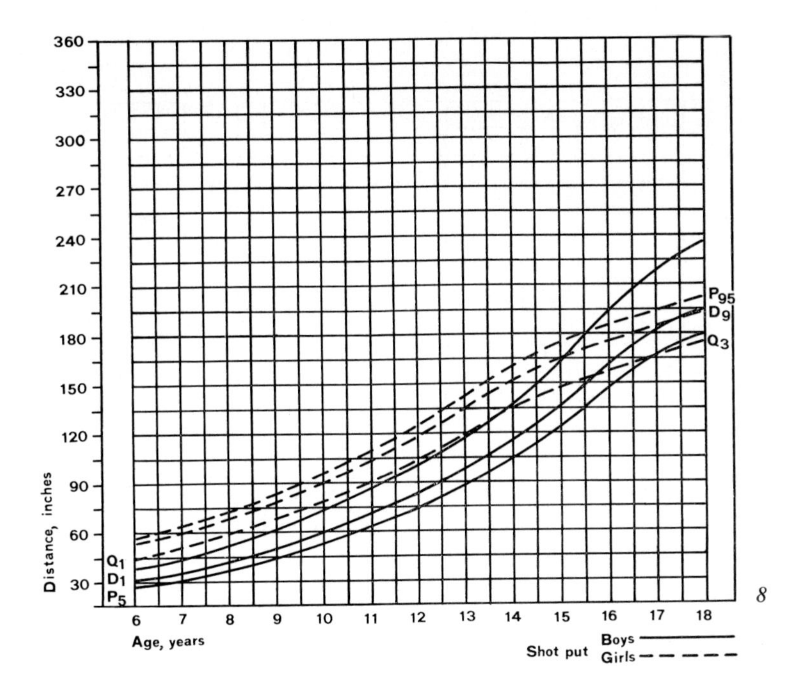

8

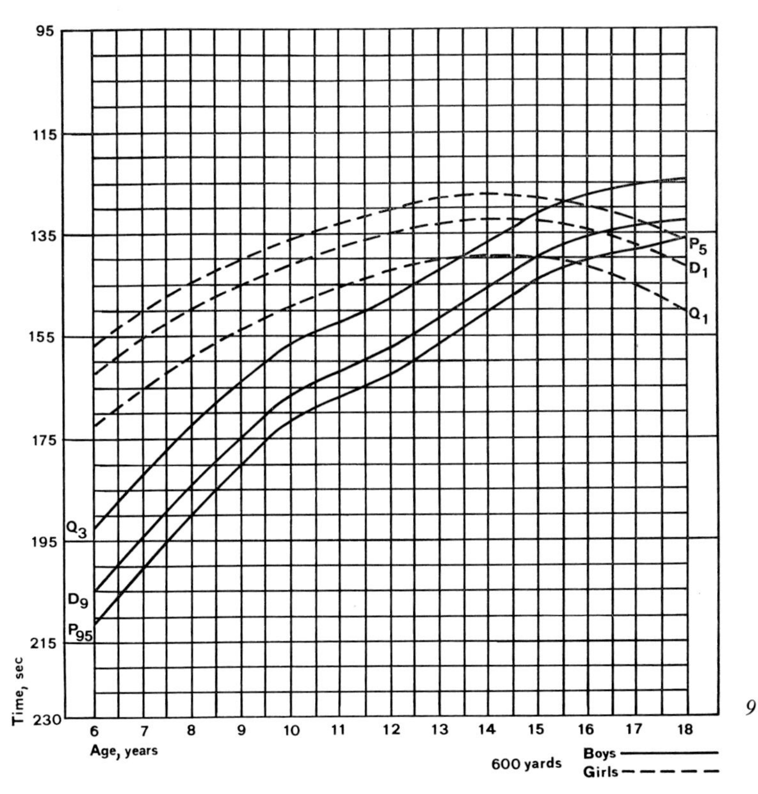

9

For legends see p. 12

10

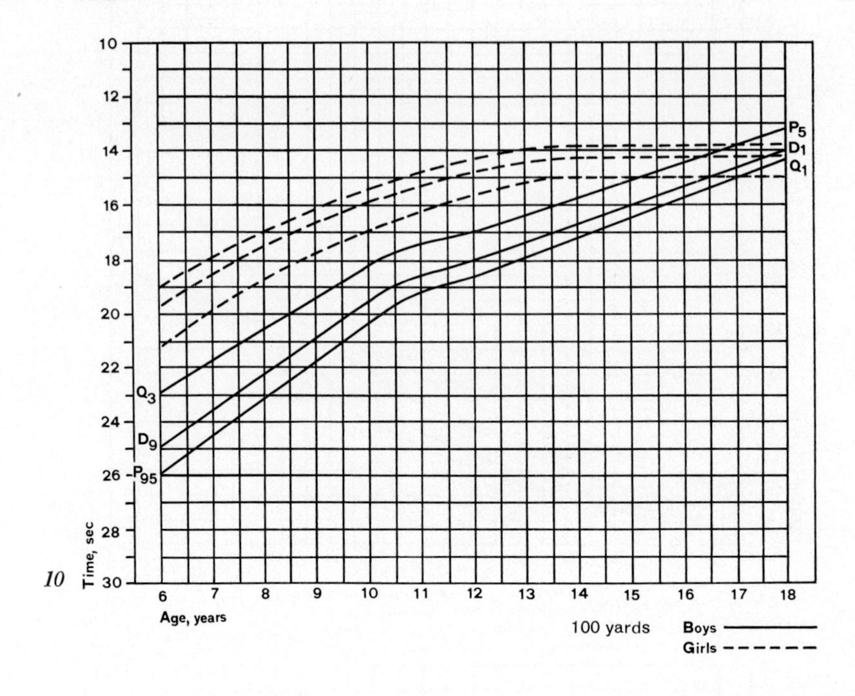

Fig. 8–10. The uppermost channels of the girls' grids are presented together with the lowermost channels of the boys' grids. All tests were conducted with untrained children.

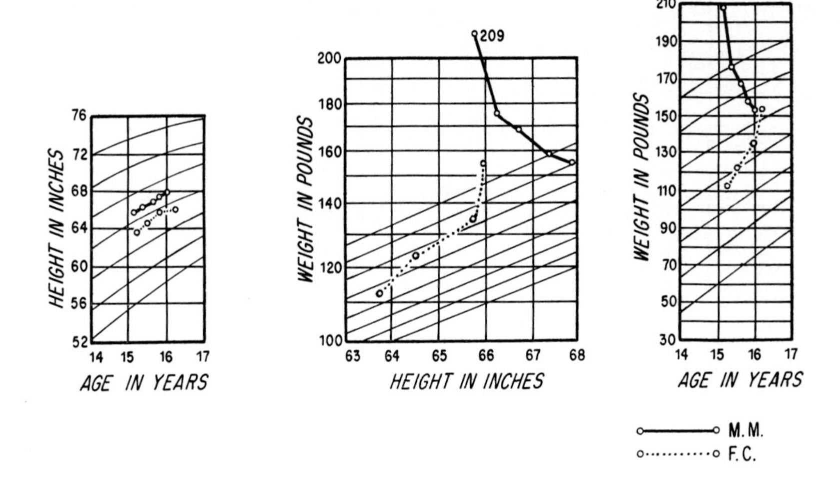

Fig. 11. Effect of intensive physical training on obese and lean boys. M.M., 15 years old, was extremely fat. His height was 5 ft. 5¾ inches, his weight 209 lb. Dysplastic features were present. He had overdeveloped breasts, marked accumulation of fat about the pelvis and thighs, and a typical mons pubis. Puberty was delayed, and he was knock-kneed. Mentally he was sluggish, suffered from headaches, and gave a dull and unhappy impression. His physical efficiency was poor. Consent of his parents was obtained to have him join a boarding school where he participated in a well-designed and expertly supervised physical training program. 3–4 h each day were devoted to calisthenic exercises, swimming, games, and track and field activities. The exercise regime proved to be most effective. Instead of gaining 12 lb., as most boys of his age do during this period of growth, he lost 55 lb. After one year his weight was within the normal range of distribution for boys of his age. Puberty had set in, he became alert, developed initiative and felt happy in the new environment. His physical efficiency had improved significantly, and he was a popular member of a successful softball team.

A corresponding opposite to the above case was that of F.C., a boy of the same age who weighed about 100 lb. less when he joined the school. His height was normal. During the ensuing year he grew 2¼ inches in line with the standard growth ascent of his channel. His weight however, increased by 43 lb., 30 lb. more than the standard ascent of his growth channel. Intensive physical training normalises body composition by increasing muscle mass and burning up fat.

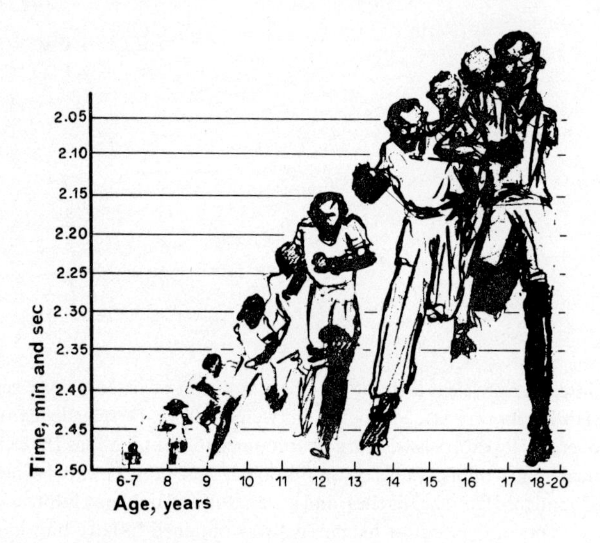

600 yards running test

Fig. 12. Drawings plotted to scale showing different growth patterns of physical endurance of girls (above) and boys (below). Performance maxima in the 600 yards running test were reached by girls at age 13; the boys' growth curve in this event ascends continuously.

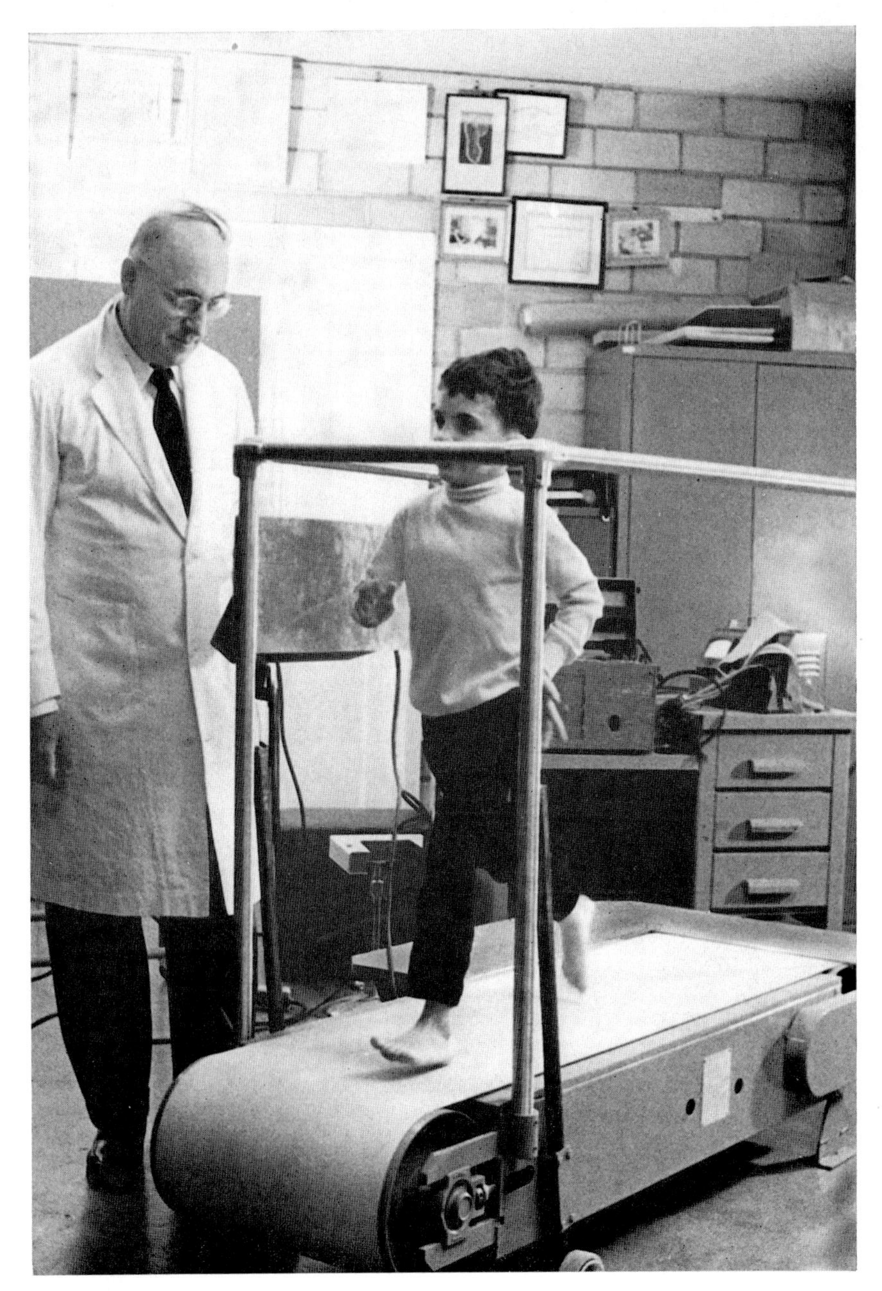

Fig. 13. 4-year-old boy running on treadmill for 15 min at a speed of 5 mph. Most young children enjoy running; their endurance is great.

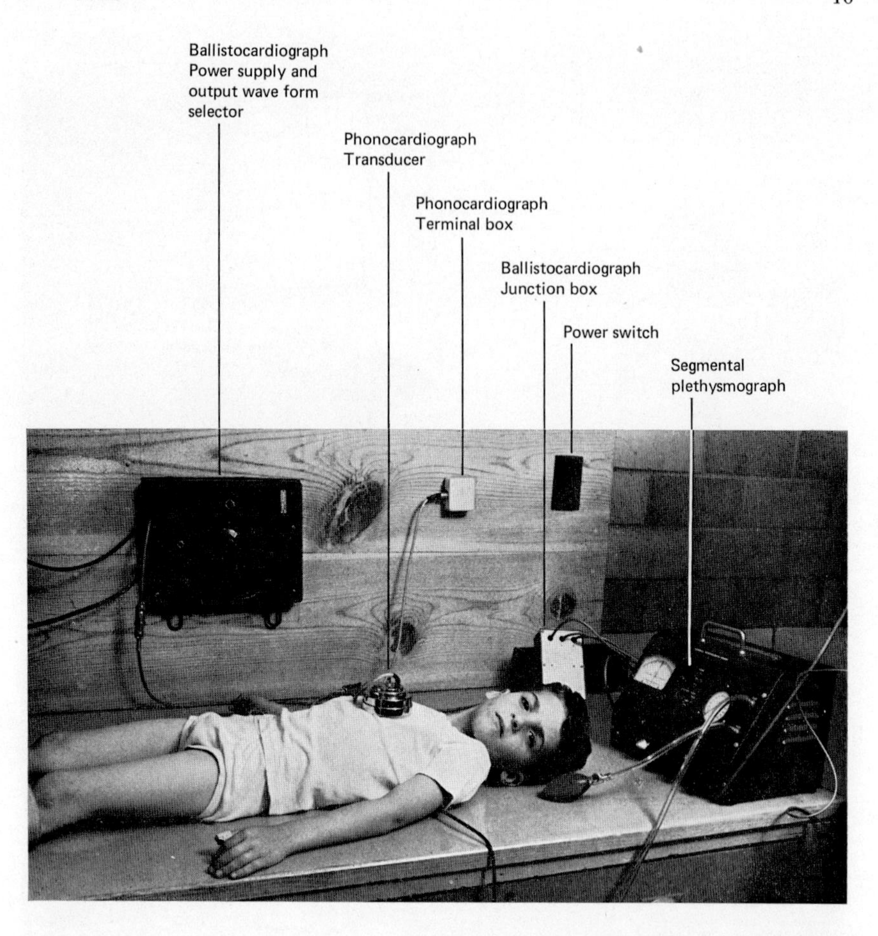

Fig. 14. Boy being tested in research laboratory. Functional parameters under investigation are indicated. Bright young children display keen interest in tests, cooperate effectively.

Fig. 15. Groups of untrained school girls aged 6, 10, 14 and 18, after 600 yards running test. The 6-year-olds are least, the 18-year-olds most fatigued.

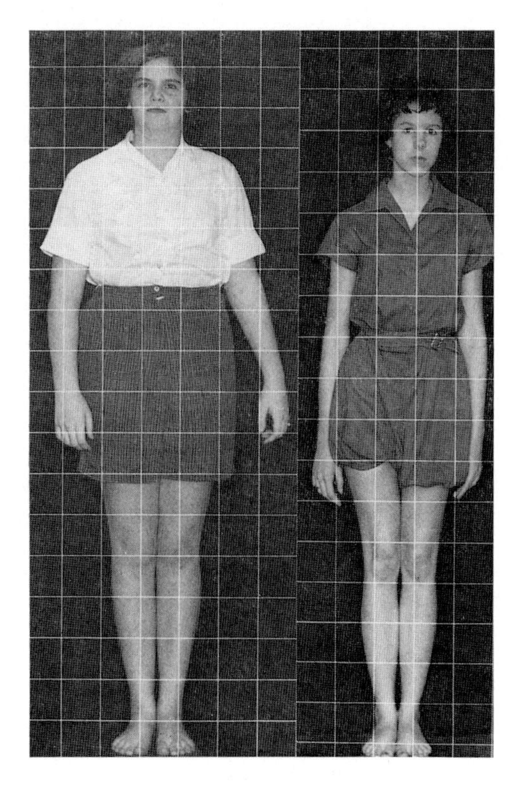

Fig. 16. Two 14-year-old girls attending the same high school in Lexington, Kentucky. The effect of different physiques upon physical performances can be identified with the help of the grids (see fig. 5–7, pp. 9, 10).

Fig. 17. The high physical performance potential of small children revealed by research in Pediatric Work Physiology is not confined to endurance. Boys and girls of pre-school age also are capable of acquiring a great variety of skills, and to develop considerable muscular strength.

Elaborate curricula of physical training for children 2–4 years of age have been prepared consisting of exercise tasks designed to fit their intellectual and emotional maturity. To illustrate this, five calisthenic exercises are presented. Due consideration was given to the fact that small children cannot appreciate the significance of a formal instruction of the kind given to adolescents and adults in physical education classes. But children do respond readily if an appeal is made to their imagination. If a small child is fond of an elephant it is not difficult to induce it 'to become an elephant'. Once it has become an elephant it will move like an elephant. In this way the children can easily be persuaded to carry out many kinds of exercises [JOKL, 1940].

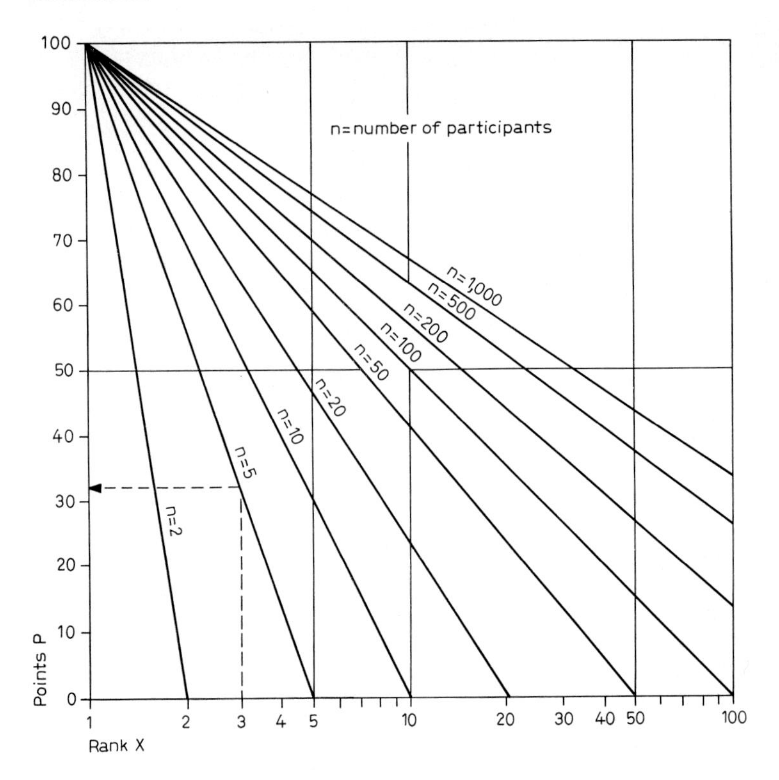

Fig. 18. The 'point allocation system' is based on the 'rank' of each competitor. The winner in any given Olympic event is awarded 100 points thus making comparison between different sports possible. The last-ranking competitor receives 0 points. The 'point distance' between the winner and the one placed second, between the second and the third, etc., depends on the total number of participants in the event. The differences in the angles of the diagonal lines are illustrated. For each possible number of participants (n) a separate line is drawn. Several samples of such lines are shown in the graph. The height of the vertical line drawn from any rank number X until crossing the appropriate n-line indicates the point value P of the rank in question. Example: five competitors (n=5); third man (X=3) receives 32 points (P = 32) [Jokl *et al.*, 1956].

Fig. 19. Drawing by GOETHE made in Weimar in 1781 showing the height of a recruit being measured. The pitiful figure for the young woman on the right suggests that enlistment was not altogether voluntary.

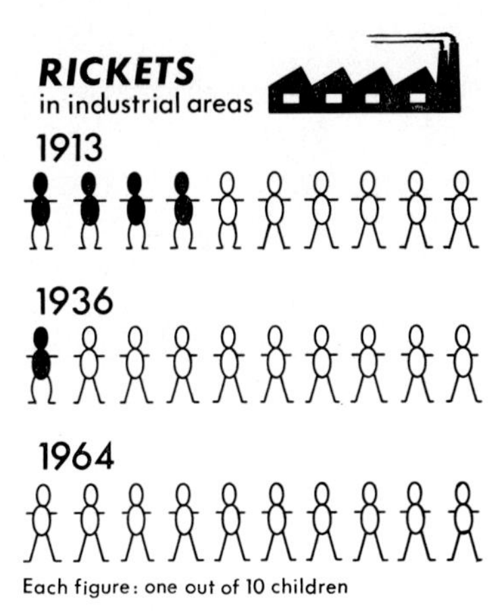

Fig. 20. In 1913 Sir JOHN BOYD-ORR assessed the incidence of rickets in Glasgow, Scotland: 40% of the city's school population were found to be afflicted. In a repeat survey in 1936 only 10% of the children in Glasgow had rickets. The study was repeated once more in 1964: not a single case of rickets was now found among children who had grown up in Glasgow. The incidence of rickets among Asian immigrants to England in 1973 was as high as it had been in Glasgow in 1913.

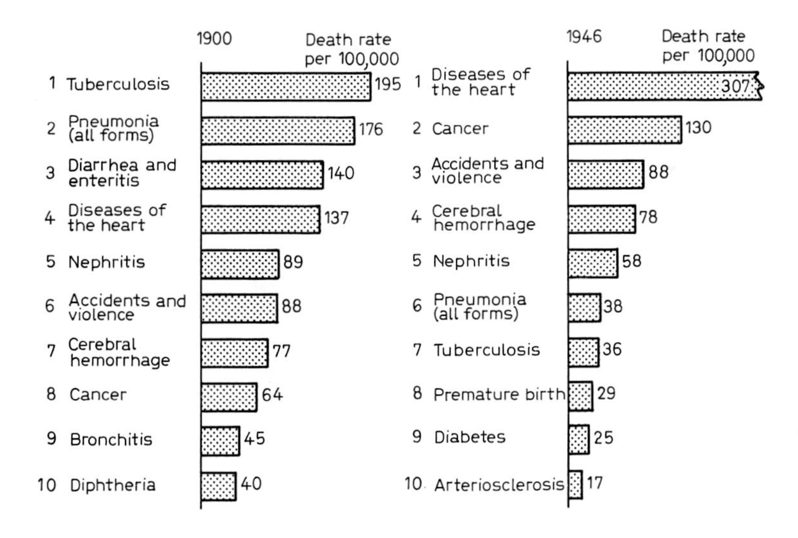

Fig. 21. Rank of the leading causes of death in the US, 1900 and 1946. In 1900 infectious diseases were responsible for most deaths in the United States. In 1946 the chief causes of death were cardiovascular diseases, malignant tumors, accidents, and violence.

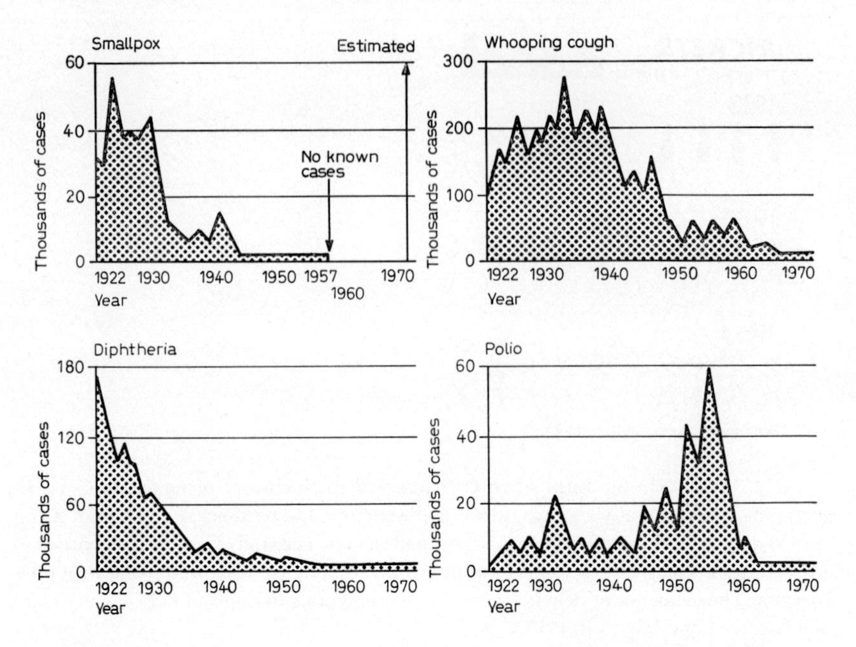

Fig. 22. Four major infectious diseases virtually eliminated in the United States by vaccines.

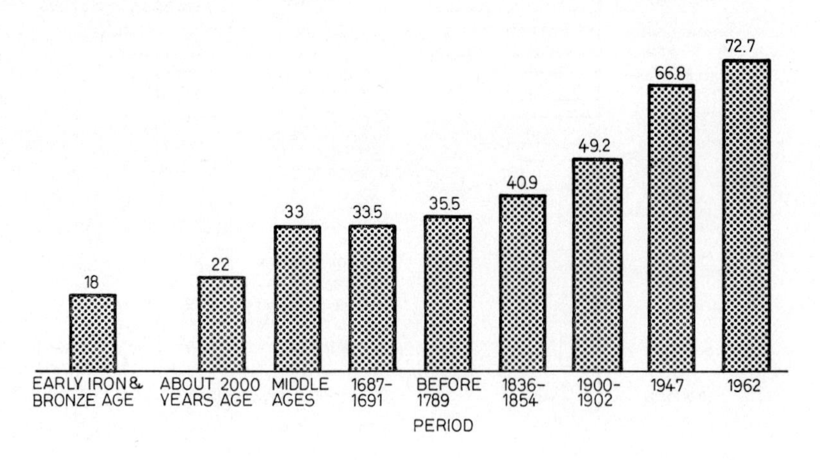

Fig. 23. Americans today live more than twice as long as people lived at the turn of the 18th century. Since 1900 an increase of longevity amounting to almost 25 years has taken place.

Table I. Shop put performance in boys and girls

Age	Number	Mean ft and inch	SD inch	SEM inch	V %	SEV %	P_5 ft and inch	D_1 ft and inch	Q_1 ft and inch	Md ft and inch	Q_3 ft and inch	D_9 ft and inch	P_{95} ft and inch
6– (and) 7– *	503	4 10.44	15.198	0.678	26.01	0.874	3 00.02	3 03.23	3 11.28	4 09.64	5 08.16	6 05.49	7 00.46
**	489	3 06.06	12.066	0.546	28.69	0.990	2 00.56	2 03.10	2 09.23	3 05.22	4 01.69	4 10.56	5 03.59
8– *	432	5 10.44	17.778	0.855	25.24	0.911	3 08.22	4 00.95	4 10.11	5 07.68	6 09.68	7 10.70	8 06.30
**	397	4 06.72	14.004	0.703	25.59	0.965	2 09.46	3 01.38	3 08.48	4 05.34	5 03.64	6 02.17	6 08.81
9– *	480	6 10.68	20.484	0.935	24.78	0.847	4 01.78	4 07.43	5 08.74	6 09.06	8 00.19	9 01.78	9 08.73
**	388	5 01.08	14.838	0.753	24.29	0.922	3 02.96	3 07.51	4 02.33	4 11.42	5 10.69	6 08.76	7 03.69
10– *	530	8 01.20	23.832	1.035	24.52	0.797	5 01.43	5 07.80	6 07.44	7 11.72	9 05.58	10 09.27	11 07.91
**	373	5 11.34	17.406	0.901	24.40	0.945	3 06.71	4 01.51	5 00.00	5 10.17	6 10.54	7 10.49	8 06.67
11– *	520	9 01.62	24.894	1.082	22.71	0.733	5 07.05	6 05.59	7 09.32	9 00.85	10 06.52	11 09.28	12 08.13
**	424	6 11.58	18.132	0.881	21.69	0.779	4 06.53	4 11.62	5 11.43	6 10.76	8 00.13	8 10.59	9 06.98
12– *	610	10 10.02	29.034	1.176	22.33	0.970	6 11.40	7 08.75	9 01.56	10 10.25	12 04.74	14 00.55	14 09.94
**	439	8 00.39	22.554	1.076	23.40	0.832	5 02.70	5 09.29	6 09.10	7 09.80	9 03.06	10 07.94	11 04.33
13– *	602	12 05.34	35.646	1.453	23.87	0.726	7 10.42	8 08.54	10 04.95	12 02.05	14 03.96	16 04.51	17 05.96
**	438	9 03.33	25.752	1.230	23.13	0.822	5 10.55	6 06.24	7 08.64	9 02.50	10 07.50	10 02.19	13 02.19
14– *	515	15 00.00	47.550	2.095	26.42	0.878	9 01.83	10 04.60	12 01.81	14 05.69	17 06.75	20 02.62	22 04.87
**	399	10 01.92	26.124	1.308	21.43	0.789	6 07.98	7 03.96	8 07.16	10 02.03	11 07.79	13 00.00	13 11.29
15– *	450	17 03.69	51.816	2.443	24.95	0.882	10 00.60	11 08.77	14 02.20	17 02.82	20 02.70	23 01.09	24 04.50
**	310	10 11.55	26.328	1.495	20.01	0.835	7 04.25	8 00.92	9 06.08	11 00.15	12 04.20	14 02.50	14 09.64
16– *	309	10 05.46	50.610	2.879	21.68	0.912	12 04.14	13 09.90	16 03.15	19 06.82	22 04.18	25 00.15	26 09.04
**	199	11 10.17	27.402	1.942	19.27	1.001	8 04.24	9 00.00	10 02.30	11 07.86	13 05.50	15 20.73	16 15.33
17 (and over) *	250	12 08.10	49.362	3.122	18.98	0.879	14 07.50	15 11.14	18 11.25	21 07.80	24 09.67	27 01.33	28 00.75
**	152	12 10.08	29.766	2.414	19.32	1.148	9 01.37	9 07.89	11 01.20	12 08.12	10 04.36	15 11.85	16 05.55

* = Boys. ** = Girls.

Table II. Time needed for boys and girls to run 600 yards

Age		Number	Mean min and sec	SD sec	SEM sec	V %	SEV %	P_5 min and sec	D_1 min and sec	Q_1 min and sec	Md min and sec	Q_3 min and sec	D_9 min and sec	P_{95} min and sec
6- (and) 7-	*	500	2 50.72	16.740	0.749	9.81	0.313	2 25.00	2 28.57	2 38.30	2 50.54	3 02.41	3 13.39	3 18.57
	**	474	3 02.85	18.245	0.838	9.98	0.327	2 33.03	2 38.48	2 49.81	3 02.86	3 15.35	3 27.04	3 33.46
8-	*	423	2 38.85	15.575	0.757	9.80	0.340	2 15.83	2 20.04	2 27.63	2 37.24	2 48.57	3 00.45	3 07.75
	*	395	2 48.25	17.370	0.874	10.32	0.371	2 22.58	2 27.08	2 34.35	2 47.37	2 59.76	4 10.16	3 17.66
9-	*	463	2 31.75	15.605	0.725	10.28	0.341	2 10.28	2 13.33	2 20.00	2 29.40	2 40.00	2 53.92	3 01.16
	**	384	2 43.95	17.440	0.890	10.64	0.388	2 18.60	2 23.57	2 31.95	2 42.00	2 54.47	3 06.53	3 17.33
10-	*	518	2 25.78	14.230	0.625	9.76	0.306	2 05.58	2 10.29	2 15.87	2 23.89	2 33.43	2 44.82	2 53.26
	**	358	2 39.20	16.010	0.846	10.06	0.380	2 16.41	2 19.61	2 27.37	2 37.27	2 49.77	3 00.46	3 07.77
11-	*	511	2 23.95	14.615	0.647	10.15	0.321	2 03.13	2 06.76	2 13.25	2 22.15	2 32.09	2 45.24	2 53.39
	**	417	2 36.55	16.050	0.786	10.25	0.358	2 13.46	2 17.16	2 25.21	2 34.61	2 45.82	3 00.00	3 07.25
12-	*	597	2 17.47	13.795	0.565	10.03	0.293	1 57.55	2 01.37	2 08.17	2 15.15	2 25.55	2 34.92	2 41.54
	**	420	2 32.50	15.800	0.771	10.36	0.361	2 08.85	2 13.46	2 21.38	2 30.94	2 42.27	2 53.81	3 00.45
13-	*	585	2 12.74	14.015	0.579	10.56	0.312	1 54.65	1 57.37	2 03.53	2 10.08	2 19.14	2 30.60	2 39.48
	**	401	2 30.45	15.430	0.771	10.26	0.366	2 06.51	2 10.87	2 19.22	2 29.32	2 41.32	2 50.47	2 55.89
14-	*	506	2 07.75	14.340	0.638	11.23	0.357	1 50.30	1 52.55	1 58.56	2 05.00	2 13.18	2 26.69	2 40.35
	**	342	2 31.45	16.175	0.875	10.68	0.413	2 08.61	2 12.20	2 19.21	2 29.05	2 42.42	2 52.84	2 58.42
15-	*	437	2 03.20	12.440	0.595	10.10	0.345	1 46.53	1 49.13	1 54.62	2 00.94	2 09.68	2 20.82	2 26.98
	**	272	2 32.38	16.485	0.999	10.82	0.470	2 07.71	2 12.00	2 20.44	2 30.30	2 42.94	2 56.27	3 03.37
16-	*	297	1 59.45	11.135	0.646	9.32	0.386	1 44.62	1 46.59	1 51.46	1 57.93	2 04.97	2 14.03	2 22.80
	**	149	2 35.10	14.470	1.185	9.33	0.545	2 08.89	2 23.31	2 23.31	2 33.75	2 46.48	2 55.92	3 02.55
17 (and over)	*	220	1 58.55	12.250	0.826	10.33	0.497	1 41.39	1 44.44	1 50.36	1 56.85	2 04.48	2 15.00	2 23.12
	**	103	2 40.15	18.535	1.826	11.57	0.817	2 07.68	2 18.30	2 26.98	2 38.65	2 51.62	3 07.12	3 12.37

* = Boys. ** = Girls.

Table III. Time needed for boys and girls to run 100 yards

Age		Number	Mean sec	SD sec	SEM sec	V %	SEV %	P_5 sec	D_1 sec	Q_1 sec	Md sec	Q_3 sec	D_9 sec	P_{95} sec
6– (and) 7–	*	503	20.45	2.453	0.109	12.00	0.384	16.96	17.51	18.75	20.08	21.86	23.99	24.94
	**	489	21.84	2.411	0.109	11.04	0.357	18.30	18.95	20.09	21.58	23.32	25.23	26.27
8–	*	432	18.79	2.006	0.097	10.68	0.367	15.75	16.34	17.47	18.51	20.00	21.42	22.58
	**	397	19.65	2.057	0.103	10.47	0.376	16.49	17.18	18.19	19.49	21.01	22.41	23.35
9–	*	480	17.78	1.781	0.081	10.02	0.326	15.33	15.70	16.47	17.57	18.32	20.18	21.17
	**	388	18.69	1.838	0.093	9.83	0.356	15.86	16.46	17.38	18.54	19.72	21.17	21.95
10–	*	530	16.84	1.525	0.066	9.06	0.280	14.62	15.05	15.76	16.66	17.69	18.97	19.97
	**	373	18.05	1.651	0.085	9.15	0.338	15.68	16.13	16.86	17.83	18.98	20.38	21.24
11–	*	529	16.39	1.427	0.062	8.71	0.270	14.34	14.68	15.32	16.23	17.28	18.36	19.10
	**	424	17.18	1.484	0.072	8.64	0.299	15.00	15.28	16.15	17.08	18.09	19.18	19.85
12–	*	610	15.77	1.328	0.054	8.42	0.243	13.72	14.13	14.89	15.60	16.58	17.50	18.24
	**	439	16.64	1.522	0.073	9.15	0.311	14.55	14.94	15.47	16.41	17.59	18.61	19.29
13–	*	602	15.29	1.284	0.052	8.40	0.244	13.48	13.75	14.36	15.13	16.04	17.09	17.73
	**	438	16.17	1.447	0.069	8.95	0.305	14.15	14.50	15.11	16.01	17.07	18.08	19.00
14–	*	515	14.65	1.301	0.057	8.88	0.270	12.73	13.10	13.73	14.50	15.33	16.37	17.12
	**	399	16.00	1.292	0.065	8.08	0.288	14.07	14.41	15.08	15.88	16.79	17.79	18.35
15–	*	450	14.13	1.270	0.060	8.99	0.302	12.32	12.68	13.25	13.95	14.85	15.70	16.42
	**	310	15.98	1.349	0.077	8.44	0.341	13.87	14.32	15.04	15.87	16.78	17.67	18.40
16	*	309	13.50	1.044	0.059	7.73	0.313	12.24	12.28	12.76	13.36	14.15	14.89	15.34
	**	199	15.86	1.328	0.094	8.37	0.432	13.75	14.15	14.84	15.83	16.77	17.51	18.17
17– (and over)	*	250	13.04	1.035	0.065	8.94	0.357	11.54	11.74	12.27	12.03	13.72	14.42	14.90
	**	152	15.81	1.261	0.102	7.98	0.460	13.85	14.22	14.83	15.76	16.63	17.54	17.97

* = Boys. ** = Girls.

Table IV. Per capita income and Olympic participation and achievement

Per capita income in dollars	Population in millions	Participations	Participation rate	Point share	Point rate	Point level
Less than 60	617	55	0.09	969	1.6	17.6
60– 99	478	105	0.22	1,825	3.8	17.4
100–199	377	609	1.62	10,566	28.0	17.3
200–499	521	1,451	2.79	41,517	79.7	28.6
500–749	113	922	8.16	22,384	198.1	24.3
750 and over	152	194	1.28	9,016	59.4	46.5

Table V. Death rate and Olympic participation and achievement

Deaths per 1000 inhabitants	Population in millions	Participations	Participation rate	Point share	Point rate	Point level
5.0– 9.9	275	802	3.28	25,962	94.4	28.8
10.0–14.9	460	1,493	3.25	35,667	77,6	23.9
15.0–19.9	389	122	0.31	1,982	5.1	16.2
20.0–24.9	205	212	1.03	3,708	18.0	17.5

Table VI. Infant mortality and Olympic participation and achievement

Deaths per 1000 live births	Population in millions	Participations	Participation rate	Point share	Point rate	Point level
25– 49	187	688	3.68	22,011	117.1	32.0
50– 74	160	621	3.88	15,564	97.2	25.1
75– 99	157	580	3.69	13,428	85.5	23.2
100–124	90	238	2.64	5,010	55.5	21.1
125–149	97	415	4.28	9,405	97.0	22.7
150–174	386	116	0.30	1,880	4.9	16.2
175–199	74	172	2.32	3,351	45.3	19.5
200–224	18	5	0.30	74	4.0	14.8

Table VII. Olympic participation and achievement of nations on different levels of caloric consumption

Kilocalories per head per day	Population in millions	Participations	Participation rate	Points	Point rate	Point level
1,500–1,999	432	40	0.09	610	1.4	15.3
2,000–2,499	749	397	0.53	8,709	11.6	21.9
2,500–2,999	82	264	3.24	5,936	72.7	22.5
3,000–3,499	232	969	4.17	29,324	126.1	30.3

Table VIII. Olympic participation and achievement of the temperature zones

Temperature zones[1]	Population in millions	Participations	Participation rate	Point share	Point rate	Point level
Cold	312	1,049	3.36	29,552	94.7	28.2
Cold and cool	246	417	1.70	14,979	60.9	35.9
Cool	401	1,395	3.48	32,472	81.0	23.3
Cool and warm	495	79	0.20	2,738	5.5	34.7
Warm	866	476	0.60	7,439	8.6	15.6
Warm and hot	64	11	0.20	146	2.3	13.3

[1] Annual isotherms of temperature zones: Cold: 0–10 °C or 32–50 °F. Cool: 10–20 °C or 50–68 °F. Warm: 20–30 °C or 68–86 °F. Hot: More than 30 °C or 86 °F.

Explanations to tables IV–VIII

Results of the analysis of performances of the 4,925 participants at the 1952 Olympic Games at Helsinki, Finland. The aim of the study was to ascertain the role played by per capita income, death rates, infant mortality, caloric consumption and temperature zones. Table IV shows correlation between per capita income, Olympic participation and achievement; table V between death rates, Olympic participation and achievement; table VI between infant mortality, Olympic participation and achievement; table VII between Olympic participation and achievement of nations on different levels of caloric consumption, and table VIII shows the correlation between Olympic participation and achievement of the temperature zones.

The term participation rate pertains to numbers of participants per million inhabitants; point share to numbers of points collected by a country; point rate to numbers of points collected per million inhabitants; and point level to average number of points per participant.

Bibliography

ARIÈS, P.: Centuries of childhood. A social history of family life (Vintage Books, New York 1922).
BERENSON, B.: Rumor and reflection (Simon & Schuster, New York 1952).
BERNARD, L.: Medicine at the court of Louis XIV. Med. Hist. *6:* 3 (1962).
DUBLIN, L.I.; LORKA, A.J., and SPIEGELMAN, M.: Length of life. A study of the life table (Ronald Press, New York 1949).
HOWELLS, J.G.: World history of psychiatry (Brunner-Mazel, New York 1975).
JOKL, E.: Physical exercises. Syllabus for South African schools (v. Schaik, Pretoria 1940).
JOKL, E.; JONGH, T.W. DE, and CLUVER, E.H.: A national Manpower survey of South Africa. The principle of physical performance grids. Manpower. South African Government Publication, vol. I, No. 1 (Sept. 1942).
JOKL, E.; KARVONEN, M.J.; KIHLBERG, J.; KOSKELA, A., and NORO, L.: Sport in the cultural pattern of the world; a study of the Olympic Games of 1952 at Helsinki (Institute of Occupational Health, Helsinki 1956).
LYND, S.: English children (William Collins, London 1942).
PREECE, M.A.: Vitamin-D deficiency among Asian immigrants to Britain. Lancet, April 28 (1973).
RICHARDSON, C.D.T.: Medical status of recruits one hundred years ago. J.R. Army Med. Corps *71/1:* (1938).
WINDHAUS, A.O.: Nobel lectures, Chemistry 1922–1941. The Nobel Foundation (Elsevier, Amsterdam 1966).
WINSLOW, C.E.A.: The conquest of epidemic disease. A chapter in the history of ideas (Princeton University Press, Princeton 1943).

Prof. E. JOKL, University of Kentucky Medical School, 340 Kingsway, *Lexington, KY 40502* (USA)

Medicine Sport, vol. 11, pp. 29–31 (Karger, Basel 1978)

Use of Treadmill and Working Capacity Assessment in Preschool Children

B. MRZENA and M. MAČEK

Laboratory for Physical Fitness Research, Faculty of Pediatrics, Charles University, Prague

The assessment of aerobic power and the evaluation of adaptation to physical load are important components of modern clinical methods of examination. There are many data available for adults and older children, but physiological data are lacking for small children from 3 to 5 years. KLIMT [1971] examined 5-year-old children on the bicycle ergometer and the treadmill, as did SILVERMAN *et al.* [1972] and MOCELLIN *et al.* [1971]. Younger children under 5 years of age have not yet been studied. The cause of lack of information for this age group probably lies in the difficulty of obtaining good cooperation, since all available methods demand a certain level of training and movement maturity.

The objective of this study is the assessment of working capacity of children at the age from 3 to 5 years. The treadmill was used since walking and running are the most natural movements at this age and do not need special training.

Methods

The treadmill allows the load to be increased by increasing either the speed or the inclination. Both techniques were used in two groups of children in the age group from 3 to 5 years. The age and weight data of both groups are presented in table I.

The children of the first group (n = 4) walked or ran for 5 min each at 3 different speeds. The first speed was 3 km/h, the second 4 km/h and the third 5 km/h. During each loading speed, the heart rate, step frequency and in the last 2 min the oxygen consumption, ventilation and other parameters were estimated with the bag method. The RQ was calculated. The mouthpiece constructed in this laboratory by VÁVRA and VANEK, with small dead and lower resistance, was used.

Table I.

Group A		Group B
Age		
3 years = 3		2
4 years = 7		3
5 years = 4		5
Boys	7	5
Girls	7	5
$\overline{X}$ weight, kg	16.6	18.5

Ten children of the second group performed at the standard speed of 4 km/h, with increasing inclination each 5 min from 0 to 5, 10 and 15°. The parameters measured were the same as in the first group.

Results

In the first group, the HR increased linearly with the speed of the treadmill. The highest HR were registrated at the speed of 5 km/h (142 ± 9.7). The corresponding oxygen consumption was 366 ± 79 ml/min or 22.06 ± 4.7 ml/kg/min. The RQ in all experiments were lower than 1.0.

The highest HR (162 ± 96/min) was measured in the second group at the inclination of 15°. Oxygen consumption was higher than in the first group: 635 ± 100 ml/min or 28.7 ± 4.84 ml/kg/min. The RQ for this kind of load was also lower than 1.0.

Discussion

The method and results of the workload test in small children is apparently determined by the degree of movement maturity and the psychological and emotional state. Appropriate test behavior must be comprehensible for children. The child has to have a feeling of a play. The greatest problem is the fear of the mouthpiece which can weaken the experiment. In spite of this the mouthpiece is used, because it gives higher accuracy than the mask. Training in the use of the mouthpiece before the experiment makes it easier. The values of heart rate and oxygen consumption were lower in both kinds of exercise at the maximal load level, as expected.

The oxygen transport capacity is considered the main limiting factor of maximal working capacity in adults and older children. Nevertheless, in these experiments in small children it was impossible to increase the loads either by higher speed, or by higher inclination. Attempts were made to increase the speed to 6 km/h and the inclination to 20°, but the children were not able to increase the step frequency and lost their balance.

In the lives of normal children, higher HR than in our experiments may be found if influenced by emotions or in spontaneously moving activity. From this point of view, our results could be considered as measures of maximal power limited by movement maturity or neuromuscular coordination, which means limited working capacity. If we compare both kinds of loading methods, the more natural for small children is to increase the inclination with a standard speed. This test is suitable for very small children and makes possible the assessment of some important parameters. The attainment of maximal values in them, as in older children, is limited by the movement maturity for this kind of exercise.

References

KLIMT, F.: Treadmill exertion in children aged five. Acta paediat. scand. *217:* suppl., p. 32 (1971).

MOCELLIN, R.; LINDEMANN, J.; RUTENFRANZ, J., and SBRESNY, W.: Determination of W_{170} and maximal oxygen uptake in children by different methods. Acta paediat. scand. *217:* suppl., p. 13 (1971).

SILVERMAN, M.; SANDRA, D., and ANDERSON, K.: Metabolic cost of treadmill exercise in children. J. appl. Physiol. *33:* 696 (1972).

Dr. B. MRZENA, Laboratory for Physical Fitness Research, Faculty of Pediatrics, Charles University, *Prague* (Czechoslovakia)

Medicine Sport, vol. 11, pp. 32–38 (Karger, Basel 1978)

Development of the Functional Capacity and Body Composition of Boy and Girl Swimmers Aged 12–15 Years

Š. Šprynarová, J. Pařízková and I. Juřinová

Research Institute, Faculty of Physical Education and Sports,
Charles University, Prague

A longitudinal investigation of boys and girls from a swimming school was done to elucidate the influence of 3 years training in this school on the somatic and functional development of boys and girls. The intensity and amount of training in girls and boys did not differ, representing 11 training sessions per week. Every year in May, there was a 2-week training camp with three training sessions a day. In July and August, training did not take place. The total amount of training and distance speed swimming was small during the first year, increasing substantially in the second year and again in the third year (fig. 1).

Data on somatic and functional development during the four examinations made at 1-year intervals were obtained on 8 boys and 10 girls. Somatic development in both groups was above the average for the Czech population (fig. 2) [Kapalín, 1967].

Maximum aerobic power was assessed during a graded work load on a horizontal treadmill [Šprynarová and Pařízková, 1965] and body composition was assessed from body density [Pařízková, 1961]. As regards body weight and total lean body mass (LBM), the boys and girls did not differ throughout the period of investigation, a difference in height was manifested only at the age of 15 where the values in boys were higher. The total body fat was greater in girls at the age of 14 years, the relative body fat at the ages of 14 and 15 years.

In maximum oxygen uptake the boys and girls differed only at the age of 15 years, while they differed in maximum oxygen uptake per kg LBM at the ages of 12, 13, and 15 years and per kg body weight at all measurements. As the measurements at the age of 14 years could be done only during the 2-week training camp, fatigue resulting from the large amount of training was

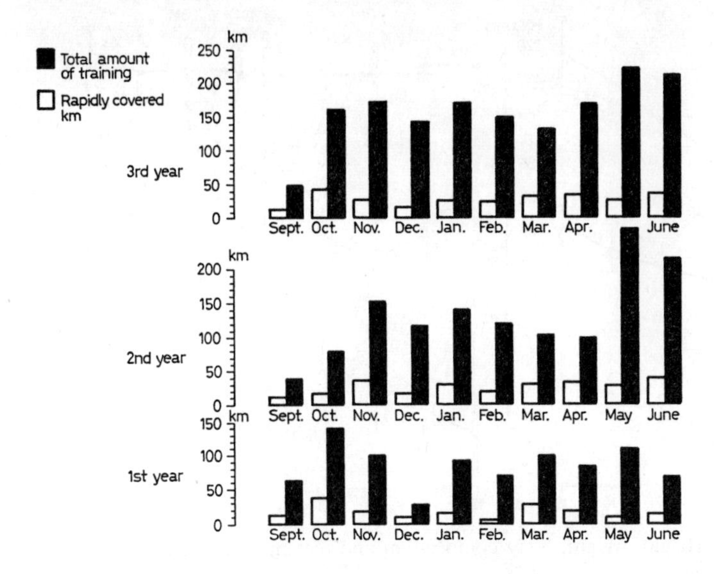

Fig. 1. The total amount of training and rapidly covered km.

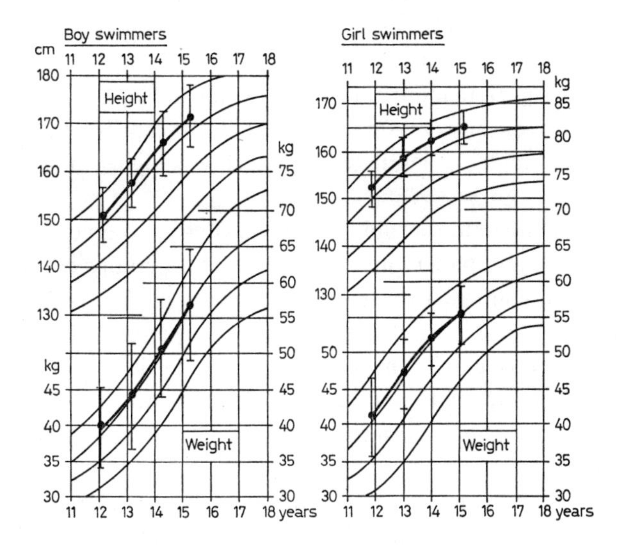

Fig. 2. Somatic development in boy and girl swimmers in the grid for Czech population.

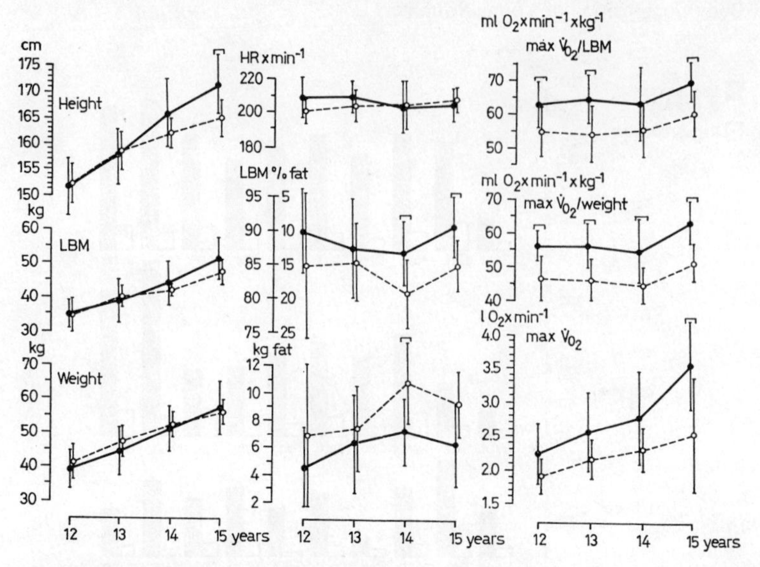

Fig. 3. Height, weight, body composition and maximum aerobic power in boy (●) and girl (○) swimmers.

manifested in the majority of boys and also in some girls by a drop of relative values of the maximum oxygen uptake. In the boys, this led to a greater scatter of maximum oxygen uptake per kg body weight and kg LBM and thus no sex differences in the values of maximum oxygen consumption per kg LBM were found in contrast with the other years of the investigation (fig. 3).

Similar patterns as in maximum aerobic power were found also in the maximum oxygen pulse (fig. 4). In indicators of ventilation during a maximum load, no differences between boys and girls were found except for the percentage of oxygen extraction from ventilated air which was higher in boys at the age of 15 years.

The development of maximum aerobic power in swimmers was compared with the development in a group of 12 untrained boys. Both groups of swimmers had participated, in contrast with the group of untrained boys, before entering the swimming school, in an average of 2 years swimming training. In spite of that, the absolute maximum oxygen uptake did not differ in all three groups at the age of 12. In relation to the body weight, it was significantly higher in the boy swimmers than in the untrained boys. The great majority of the individual values of maximum oxygen uptake per kg body weight at the age of 12 and 15 years is on the left of the line of identity and it may thus be said that the increase of functional capacity in the investigated groups between 12 and 15 years was greater than the increase of body weight (fig. 5).

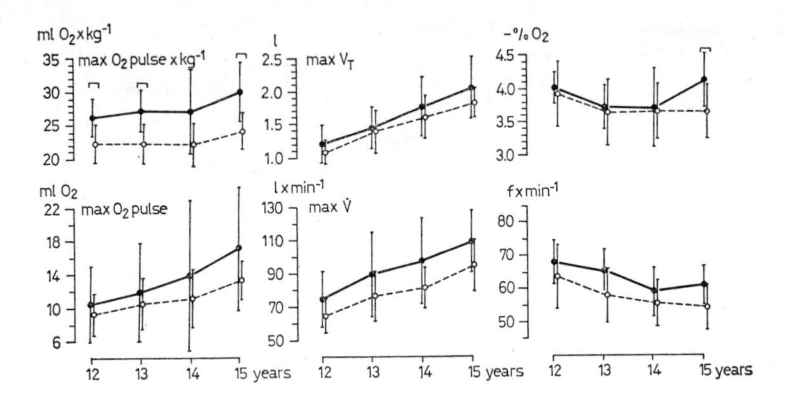

Fig. 4. Maximum O_2 pulse, maximum values of pulmonary ventilation, tidal volume, respiratory frequency and percentage of oxygen extraction from ventilated air during maximum load in boy (●) and girl (○) swimmers.

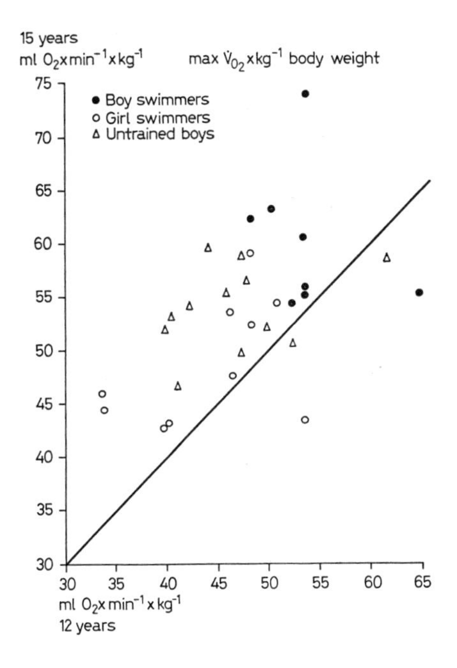

Fig. 5. The development of maximum aerobic power between 12 and 15 years of age in boy and girl swimmers and in untrained boys.

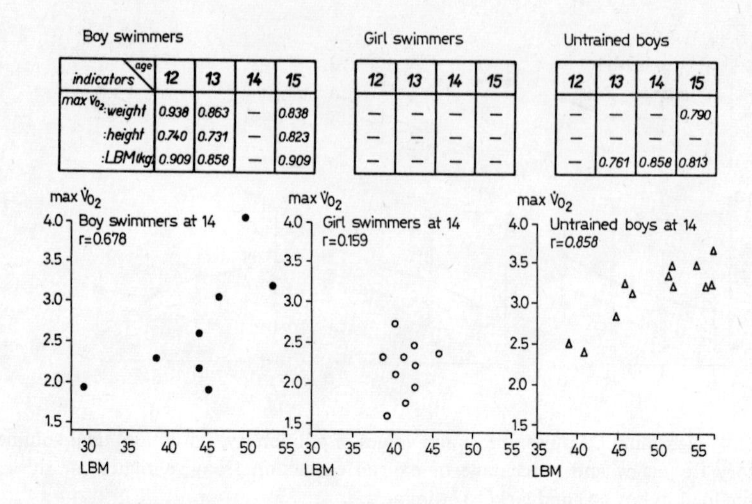
Boy swimmers

indicators \ age	12	13	14	15
max $\dot{V}_{O_2}$: weight	0.938	0.863	—	0.838
:height	0.740	0.731	—	0.823
:LBM(kg)	0.909	0.858	—	0.909

Girl swimmers

12	13	14	15
—	—	—	—
—	—	—	—
—	—	—	—

Untrained boys

12	13	14	15
—	—	—	0.790
—	—	—	—
—	0.761	0.858	0.813

Fig. 6. Significant correlations between maximum aerobic power and body dimensions in boy and girl swimmers and in untrained boys.

The results of cross-sectional surveys indicate that the maximum oxygen uptake per kg in girls declines after the age of 12 [Åstrand, 1952; Cerretelli *et al.*, 1963]. In these girl swimmers, who were subjected to intense training, on the other hand, a significant increase of this indicator occurred. Thus, in girls, it is possible to prevent by training a drop of relative values of maximum oxygen uptake after the age of 12 years. The course of changes of the maximum aerobic power in girls thus approached the course of changes recorded in boys.

Correlation analysis of the relationship of the maximum aerobic power and body dimensions in different years of the investigation revealed significant relationship in boy swimmers in all measurements with the exception of those at the age of 14. In girl swimmers, no significant relationship in any of the measurements were found and in untrained boys the only marked relationship found was between maximum oxygen uptake and the amount of LBM (fig. 6).

The above finding cannot be interpreted that the well-known relationship between the maximum oxygen consumption and body dimensions does not apply in girl swimmers and untrained boys. The causes why this relationship is not manifested in investigations of some groups may be twofold. They depend, on the one hand, on the homogeneity of the group in one or both of

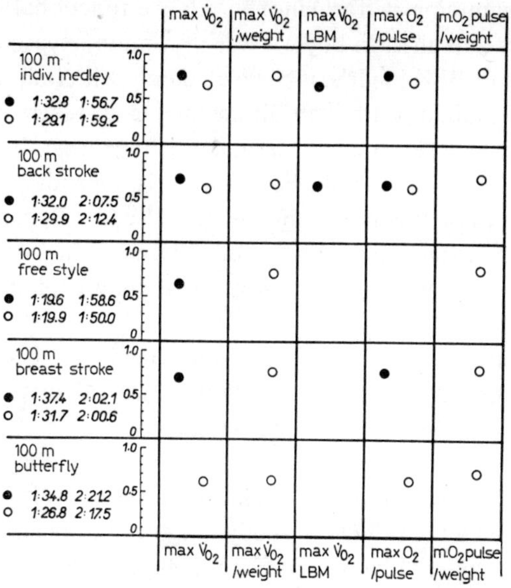

	max $\dot{V}_{O_2}$	max $\dot{V}_{O_2}$ /weight	max $\dot{V}_{O_2}$ LBM	max O_2 /pulse	m.O_2 pulse /weight
100 m indiv. medley; ● 1:32.8 1:56.7; ○ 1:29.1 1:59.2					
100 m back stroke; ● 1:32.0 2:07.5; ○ 1:29.9 2:12.4					
100 m free style; ● 1:19.6 1:58.6; ○ 1:19.9 1:50.0					
100 m breast stroke; ● 1:37.4 2:02.1; ○ 1:31.7 2:00.6					
100 m butterfly; ● 1:34.8 2:21.2; ○ 1:26.8 2:17.5					

Fig. 7. The relationship of swimming performance to functional indicators in boy (●) and girl (○) swimmers at the age of 12.

the mutually correlated indicators. The greater the homogeneity of the group in the correlated indicators, the greater the accumulation of values in a small portion of the regression line for the analyzed relationship and the smaller the correlation coefficient. The second type of causes is associated with the impact of some further factor which has a bearing on one of the correlated items and which does not act uniformly in the investigated group. The uneven action of such a further factor, e. g. a different degree of fatigue in the investigated boys aged 14, then masks the correlation.

The performance of the investigated swimmers was tested in competitions which, with the exception of the first year, were done at a time interval of more than 3 months from the functional examination. Therefore, it was possible to investigate the relationship between swimming performance and functional capacity only in 12-year-old children. In the same distance, boys and girls at this age had approximately the same times, performance became differentiated particularly at the ages of 13 and 14 years. The coefficients of rank correlation which were used to investigate the relationship between the achieved time and functional indicators were significant in boys in 4 of 5 tested disciplines, in girls in all 5 disciplines.

The relationship of swimming performance to relative functional indicators was different in boys and girls. In boys, it was maximum oxygen consumption in relation only to LBM; in girls, it correlated only with indicators in relation to kg of body weight. Although all swimming distances were 100 m and the performance lasted less than 2.5 min, it was found that the performance on these distances in 12-year-old boys and girls depends on the maximum oxygen uptake and on the maximum oxygen pulse (fig. 7). A more rapid adaptation of the circulation and respiration of children to the load and their lower anaerobic capacity as compared with adults may explain this finding.

It is difficult to explain why in boys a relationship was found between performance and relative functional indicators only in values calculated per kg LBM and in girls only in values calculated per kg weight.

As a certain amount of fat is no impediment for swimming, perhaps the ratio of body fat in total body weight in girls could play a certain role in the relationship of performance and max VO_2 per body weight. But the fact that girl and boy swimmers at the age of 12 did not differ as regards body weight or body composition suggest evidence against this interpretation. Obviously, it would be necessary to compare greater groups of boy and girl swimmers to explain these different relationships, something not possible in the present study.

References

Åstrand, P. O.: Experimental studies of physical working capacity in relation to sex and age (Munksgaard, Copenhagen 1952).

Cerretelli, P.; Aghemo, P., and Rovelli, E.: Morphological and physiological observations on school children in Milan. Med. Sport. *3:* 109–121 (1963).

Kapalín, V.: Lékařské repetitorium. Medical handbook, pp. 109–117 (Charvát, Praha 1967).

Pařízková, J.: Age trends in fatness in normal and obese children J. appl. Physiol. *16:* 173–174 (1961).

Šprynarová, Š. and Pařízková, J.: Changes in the aerobic capacity and body composition in obese boys after reduction. J. appl. Physiol. *20:* 934–937 (1965).

Dr. Š. Šprynarová, Research Institute, Faculty of Physical Education and Sports, Charles University, *Prague* (Czechoslovakia)

Medicine Sport, vol. 11, pp. 39–46 (Karger, Basel 1978)

Muscle Blood Flow in Prepubertal Boys

Effect of Growth Combined with Intensive Physical Training

GÜNTER KOCH

Department of Clinical Physiology, Central Hospital, Karlskrona

While information regarding the adaptation of central hemodynamics and pulmonary gas exchange to exercise is rapidly increasing [ERIKSSON *et al.*, 1971; ERIKSSON and KOCH, 1973; KOCH and ERIKSSON, 1973a, b], there are very few data available concerning the adjustment of the peripheral circulation. Some data concerning the magnitude and adjustment of muscle blood flow (MBF) in the working muscle of the leg in prepubertal boys have recently been presented [KOCH, 1974].

The present paper reports the changing pattern of MBF observed in the prepubertal age group while growing older from 12 to 13 years under the influence of intensive physical training.

Material

Nine boys volunteered with their parents' consent for the study. At the time of the initial study in 1973 they were 11.9 ± 0.2 (range 11.8–12.3) years of age, 150 ± 4.3 (range 145–159.5) cm tall and weighed 39.2 ± 4.0 (range 31.0–43.5) kg. They were re-studied after exactly 12 months (1974), at that time having a body height of 156 ± 5 (range 151–165) cm and a body weight of 43.2 ± 4.1 (range 34.7–49.5) kg.

Prior to the initial study, all the boys were highly motivated by a very strong interest in physical exercise and sport, regularly took part in different sporting activities, such as football, athletics, ice hockey, etc., and had a high maximal oxygen uptake (mean 59.5 ml/kg BW). After the first study in 1973, they were encouraged to continue at the same level or, preferably, increase physical training. An evaluation of the intensity of physical activity performed during a 40-week period between the two studies is given in table I.

Table I. Evaluation of intensity of physical activity during last 40-week period

	Mean	SD	Min	Max
Total number days/week with physical exercise	4.61	0.59	3.78	5.28
Number days/week with running exercise (average 3–5 km/session)	1.73	1.22	0.43	4.45
Number days/year reported sick at school	16.1	17.1	0	50

In addition: 3 h/week school physical education (mainly ball games).

Methods

On both occasions, in April, 1973, and April, 1974, the boys were studied with identical techniques and under identical conditions, the sequence of the different examination routines being exactly the same. MBF was measured with the 133-xenon clearance method, i.e. by determination of the clearance rate of 133-xenon after injection into the muscle to be studied; for details, see CLAUSEN and LASSEN [1971].

Blood flow was measured in two different situations on two consecutive days:

(1) After 'ischemic' work in the tibialis anterior muscle, i.e. the subject bending and extending his foot against a resistance with a pressure cuff inflated to more than 200 mm Hg applied around the thigh. Under these circumstances, work of 2–3 min duration can be sustained. Maximal MBF occurs immediately after release of the cuff pressure concomitant with the pronounced hyperemia induced by local hypoxia and acidosis.

(2) In the vastus lateralis muscle during successive submaximal and maximal exercise on a bicycle ergometer, the loads having been determined during a previous exercise test.

Details concerning the measuring procedure, calculations and the equipment used have been previously reported [KOCH, 1974].

Prior to the measurement of MBF, the boys had undergone a submaximal (loads 250, 500 and 750 kpm/min) and a maximal exercise test in the same manner as previously described [ERIKSSON and KOCH, 1973].

Prior to the determination of post-ischemia MBF, oscillometry was performed over the midportion of the thigh and the calf.

Table II. Pulmonary ventilation, oxygen uptake and blood lactate (means and standard deviations) at rest and during submaximal work ($W_{170} = 656 \pm 110$ kpm/m) as obtained at re-study (1974) and mean changes ($\overline{D}$) compared with initial study

	Rest			750 kpm/m		
	M	(SD)	$\overline{D}$	M	(SD)	$\overline{D}$
$\dot{V}_E$, liters	17.8	(3.2)	+3.1	71.3	(8.8)	−0.1
$\dot{V}_{O_2}$, liters	0.32	(0.04)	+0.08*	1.75	(0.18)	+0.21*
$\dot{V}_E/V_{O_2}$	56.4	(6.9)	−6.9	40.8	(2.6)	−4.3*
R	0.87	(0.12)	−0.07	0.85	(0.88)	−0.02
Lactate, mmol/l	1.6	(0.3)	−0.6*	4.0	(1.3)	−1.3*
				4.8	(1.2)	−0.6

* $p \leqslant 0.05$.

Table III. Means and standard deviations (SD) of some circulatory and respiratory data obtained during maximal exercise (1,200–1,500 kpm/min) at re-study and mean changes ($\overline{D}$) compared with initial study

	Mean	SD	$\overline{D}$
Work load, kpm/min	1,344	88	305
Pulse rate	200	6	3
Resp. rate	53	8	3
Syst. BP, mm Hg	172	8	12
Diast. BP, mm Hg	81	3	5
Lactate (max. load), mmol/l	6.3	1.3	−0.7
Lactate (4 min a.w.), mmol/l	8.8	1.4	−0.5
$\dot{V}_E$, liters	120.7	17.6	13.6
$\dot{V}_{O_2}$, ml	2,464	246	136
$\dot{V}_{O_2}$/BW, ml	56.8	9.7	−2.8
$\dot{V}_E/\dot{V}_{O_2}$	49.7	4.0	2.6

Results

Submaximal and Maximal Exercise

Some relevant results obtained during the submaximal and maximal exercise test on the second examination (April, 1974) and changes in comparison to the first examination (April, 1973) are given in tables II and III.

W_{170} calculated from the pulse response during *submaximal steady state work* was 656 ± 110 kpm/min at the second versus 636 ± 149 kpm/min at the first study; there were no statistically significant changes in pulse rate, respiratory rate, systolic and diastolic blood pressures at rest, during or after exercise. Oxygen uptake was slightly ($p<0.05$) higher at rest and during exercise at a work load of 750 kpm/min, while blood lactate was slightly ($p<0.05$) lower (table II).

Maximal work load was $1,344 \pm 88$ kpm/min in 1974 versus $1,015 \pm 173$ kpm/min in 1973 ($p<0.001$); in spite of that, maximal $\dot{V}_{O_2}$ was only slightly higher, this difference being at the borderline of statistical significance. Due to the 10% increase of body weight, the relative maximal $\dot{V}_{O_2}$ decreased from 59.5 ± 6.1 to 56.8 ± 9.7 ml/kg ($p > 0.1$, table III).

MBF and Oscillometry

Values obtained for MBF at the re-study during ergometer bicycle exercise in the lateral vastus muscle and after ischemic work in the anterior tibial muscle, as well as changes observed in comparison with the initial study in 1973, are listed in table IV.

Maximal MBF during bicycle ergometer work (average of values at submaximal and maximal exercise for both legs) was about 75 ml/100 g/min in 1974 versus 76 ml/100 g/min in 1973. There were no significant differences in MBF between the two study occasions either at submaximal and maximal load or after exercise. Neither was there a statistically significant difference between the two legs. However, on the first study, there was a statistically

Table IV. Muscle blood flow (ml/min/100 g; means and standard deviations) in the lateral vastus muscle during bicycle ergometer exercise and in the anterior tibial muscle at rest in the supine position and during maximal hyperemia (after ischemic work) as obtained at re-study (1974) and mean changes ($\overline{D}$) compared with initial study (1973)

	Right leg			Left leg		
	mean	(SD)	$\overline{D}$	mean	(SD)	$\overline{D}$
M. vastus lat.						
Submax. work	79.4	(30.8)	18.3	78.0	(24.0)	12.1
Maximal work	70.0 } **	(29.3)	1.9	70.9 } **	(24.6)	− 3.2
After work	20.1	(14.0)	−21.9	26.7	(23.3)	−18.2
M. tibialis ant.						
Rest	9.2	(6.0)	1.5	9.7	(6.0)	1.0
Ischemic work	82.9	(37.8)	−35.3*	86.5	(33.9)	−18.9

* $p \leq 0.05$; ** $p \leq 0.01$.

significant ($p<0.05$ and $p<0.01$) increase of MBF from submaximal to maximal exercise; this difference had vanished at the re-study, maximal MBF then being attained at submaximal exercise.

Maximal MBF after ischemic work was 85 ml/100 g/min (average of both legs) in 1974, this value being about 20% lower than the values obtained in 1973. This decrease was particularly marked in the right leg ($p<0.05$). The ratio of maximal MBF during ergometer work and of MBF during ischemic work was 0.67 ± 0.24 in 1973 and had increased to 0.84 ± 0.19 in 1974 ($0.1 > p > 0.05$).

The oscillometric amplitudes recorded over the thigh and calf before and after work with each foot tended for both legs to be slightly higher in 1974 compared with 1973, the difference being most marked over the left calf (9.3 versus 7.7 mm, $p<0.01$).

Discussion

The boys were studied with identical technique at approximately 12 years and again at 13 years of age; during this year, they had gained 6.5 cm in height and 4.1 kg in weight, and had continued to have rather intensive physical training.

The main features revealed by the initial study at the age of 12 years can be summarized as follows [KOCH, 1974]:

(1) MBF after ischemic work was significantly higher than during maximal bicycle work; both were significantly higher than in the young adult. Since cardiac output tends to be somewhat lower and the arterio-venous oxygen difference somewhat higher at comparable pulse rates in this age group, the higher MBF per 100 ml of tissue would suggest that a smaller proportion of cardiac output is distributed to the working muscle, but that local flow and oxygen extraction are higher. This could be explained by a smaller effective muscle mass in these boys compared with young men.

(2) Just as in the normal adult [CLAUSEN and LASSEN, 1971], MBF reached a steady state within 30 sec after the change from one load to another, i.e. earlier than steady state is achieved with respect to pulse rate and cardiac output. This discrepancy strongly suggests that the increase of cardiac output prevailing after the first minute of exercise is due to flow increases in other than working muscles, presumably mainly in the skin for purpose of heat dissipation.

(3) Unlike the condition in the adult [CLAUSEN and LASSEN, 1971;

Nordenfelt, 1974], MBF continued to increase from submaximal to maximal exercise, maximal flow occurring only at maximal exercise.

(4) The ratio between maximal MBF during ergometer work and during ischemic work was lower in these boys than in the young adult. If one considers this ratio as a measure of the MBF actually used in relation to the total flow reserve capacity, this implies that these boys used a smaller proportion of their total reserve capacity during maximal work.

When re-studied at the age of 13 years, some significant changes in MBF were observed consisting in an adjustment towards the adult pattern [Clausen and Lassen, 1971; Nordenfelt, 1974]: maximal MBF during ergometer work was reached at submaximal work loads and the proportion of the total MBF reserve capacity used during ergometer bicycle work tended to increase, approaching the ratio observed in the adult. However, the absolute magnitude of maximal MBF both during ergometer work and after ischemic work still tended to be slightly higher than in the adult.

When evaluating the changes in MBF pattern observed, both normal growth and physical training must be taken into consideration as possible causes. The effect of these two different mechanisms was obviously of varying importance in the different individuals. Thus, one of the boys increased body dimensions from 145 cm and 31 kg to 151 cm and 34.7 kg. He had the highest training intensity and increased his relative maximal $\dot{V}_{O_2}$ from 61.7 to 74.4 ml/kg while maximal MBF during ergometer work was unchanged and MBF after ischemic work about 30% lower. On the other hand, no less than 6 boys out of 9 decreased their relative maximal $\dot{V}_{O_2}$.

The lack of significant changes in maximal MBF during ergometer work in these boys during one year of extensive physical activity is hardly surprising in view of the results obtained in the adult. Thus, it appears well established that normally the blood flow required by the skeletal muscle during exercise at identical work loads decreases after training [Stenberg, 1971] while the density of mitochondria [Holloszy, 1967; Kiessling *et al.*, 1971] and the enzyme capacity [Morgan *et al.*, 1971] increase. On the other hand, the capillary density of human skeletal muscle has been shown not to be influenced by bed rest and training [Saltin *et al.*, 1968], and no difference could be demonstrated in the capillary density and intramuscular diffusion distance between such different groups as untrained students and well-trained endurance athletes [Hermansen, 1973].

The lack of effect on MBF in this group of well-trained boys does not necessarily imply that physical training may not influence MBF in young individuals with a low degree of physical activity or marked inactivity.

The positive effect of physical training on local MBF in adult patients with decreased MBF due to obstructive arterial disease is well documented [ALPERT *et al.*, 1969].

From the ergometer work/ischemic work MBF ratio, it may be concluded that local blood flow in the working muscle does not limit the local maximum oxygen uptake.

Acknowledgement

The important contribution to this study made by AKE MARTINSSON, director of physical education, who supervised the physical exercise, is greatfully acknowledged.

This study was supported by a grant from the Swedish First of May Flower Annual Campaign for Children's Health.

References

ALPERT, J.S.; LARSEN, O.A., and LASSEN, N.A.: Exercise and intermittent claudication. Blood flow in the calf muscle during walking studied by the xenon-133 clearance method. Circulation *39:* 353 (1969).

CLAUSEN, J.P. and LASSEN, N.A.: Muscle blood flow during exercise in normal man studied by the 133-xenon clearance method. Cardiovasc. Res. *5:* 245 (1971).

ERIKSSON, B.O.; GRIMBY, G., and SALTIN, B.: Cardiac output and arterial blood gases during exercise in pubertal boys. J. appl. Physiol. *31:* 348 (1971).

ERIKSSON, B.O. and KOCH, G.: Effect of physical training on the hemodynamic response during submaximal and maximal exercise in 11–13 year old boys. Acta physiol. scand. *87:* 27 (1973).

HERMANSEN, L.: Oxygen transport during exercise in human subjects. Acta physiol. scand. suppl. 399 (1973).

HOLLOSZY, J.O.: Biochemical adaptations in muscle. Effects of exercise on mitochondrial oxygen uptake and respiratory enzyme activity in skeletal muscle. J. biol. Chem. *242:* 2278 (1967).

KIESSLING, K.H.; PIEHL, K., and LUNDQVIST, C.G.: Number and size of skeletal muscle mitochondria in trained sedentary men; in LARSEN and MALMBORG Coronary heart disease and physical fitness, p. 143 (Munksgaard, Copenhagen 1971).

KOCH, G.: Muscle blood flow after ischemic work and during bicycle ergometer work in boys aged 12 years, in BORMS and HEBBELINCK Children and exercise, p. 29, Acta paediat. Belg. (1974).

KOCH, G. and ERIKSSON, B.O.: Effect of physical training on pulmonary ventilation and gas exchange during submaximal and maximal work in boys aged 11 to 13 years. Scand. J. clin. Lab. Invest. *31:* 87 (1973a).

KOCH, G. and ERIKSSON, B.O.: Effect of physical training on anatomical R-L shunt at rest and pulmonary diffusing capacity during near-maximal exercise in boys 11–13 years old. Scand. J. clin. Lab. Invest. *31:* 95 (1973b).

MORGAN, T.E.; COBB, L.A.; SHORT, F.A.; ROSS, R., and GUNN, D.R.: Effects of long term exercice on human muscle mitochondria; in PERNOW and SALTIN Advances in experimental medicine and biology, vol. 11, p. 87 (1971).

NORDENFELT, I.: Blood flow of working muscles during autonomic blockade of the heart. Cardiovasc. Res. *8* (1974).

SALTIN, B.; BLOMQVIST, G.; MITCHELL, I.H.; JOHNSON, R.L.; WILDENTHAL, K., jr., and CHAPMAN, C.B.: Response to exercise after bed rest and after training. Circulation suppl. 7 (1968).

STENBERG, J.: Muscle blood flow during exercise. Effects of training; in LARSEN and MALMBORG Coronary heart disease and physical fitness, p. 80 (Munksgaard, Copenhagen 1971).

Dr. G. KOCH, Department of Clinical Physiology, Central Hospital, *Karlskrona* (Sweden)

Medicine Sport, vol. 11, pp. 47–51 (Karger, Basel 1978)

Automatically Controlled Ergometer for Pulse-Conducted Exercise Test

I. Välimäki, M.-L. Petäjoki, M. Arstila, P. Viherä and H. Wendelin
Cardiorespiratory Research Unit, University of Turku, Turku

In spite of reasonable success in the development of ergometric exercise tests, several difficulties are met when paediatric subjects are studied. (1) In multistage steady-state tests, the choice of sequential work loads is empirical and rather arbitrary in growing individuals. (2) The steady state itself is questionable at higher work levels. (3) It is fairly inaccurate to predict the maximal working capacity on the basis of submaximal loads. (4) Estimation of maximal aerobic power presumes maximal exertion of a child who may be entirely untrained, psychologically unmotivated or disabled by some disease. This is why the present research project was conducted to apply the pulse-conducted triangular exercise test, PCXT, by Arstila [1] in children. In this test, the braking power of a bicycle ergometer is adjusted in such a fashion that the heart rate of the subject is continuously increased at a rate of 5 beats/min^2. In other words, the first derivative of the heart rate is held constant. The philosophy of this test has been reported in earlier communications [1, 4] Previously, a control unit for a semi-automatic PCXT procedure was described and tested [2, 4]. The present technical report depicts an equipment for an entirely automated PCXT.

Method and Test Subjects

An electrically braked bicycle ergometer, Model 380, and electrocardiograph (Elema, Sweden) together with an Automatic Control Unit (ACU), constructed in this laboratory, formed the test equipment (fig. 1). The braking power of this ergometer is linearly related to the control voltage. This model is especially suited for all kinds of programmed control processes. The purpose of the ACU was to automatically accelerate the subjects heart rate by 5 beats/min in every minute starting from resting level and continuing until ex-

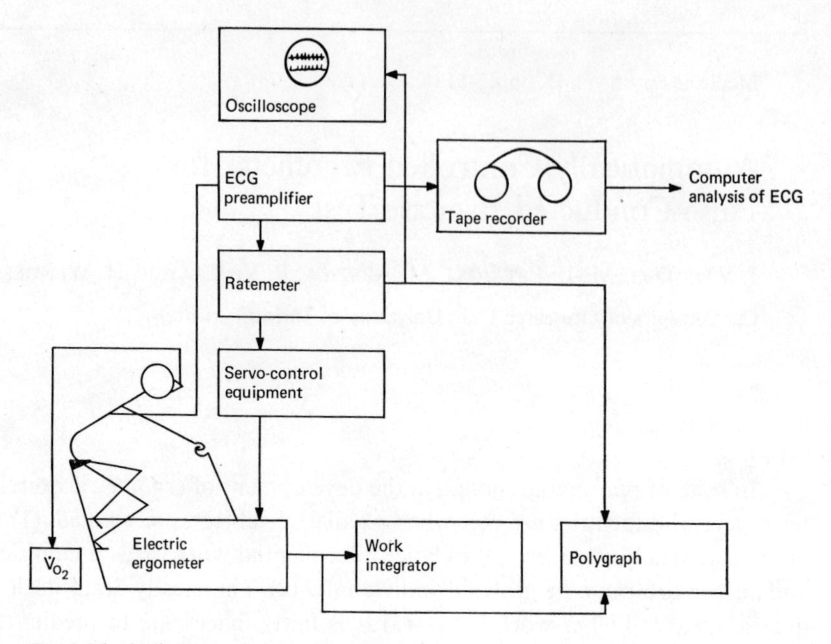

Fig. 1. Block diagram of the equipment.

haustion. The ECG of the subject was continuously monitored and a voltage (V_{HR}) corresponding to the subject's instantaneous heart rate was derived by the ACU. A voltage for reference rate (V_{REF}) of 5 beats/min^2 (optionally 0 or 4 or 8 beats/min^2) was generated by the ACU. A combination of 4 terms, utilizing these two parameters, was used in the final control process (fig. 2A): (1) a constant increase related linearly to the time; (2) the first time derivate of V_{HR}; (3) the difference V_{REF}–V_{HR}, and (4) the first time integral of V_{REF}–V_{HR}. The weighing coefficients, k_1 to k_4, of these components could be individually adjusted to find an optimal combination for the control process.

The heart rate and the braking power were presented in graphic form (fig. 2). The readings of instantaneous heart rate, difference between the heart rate and reference rate, and the total work done could be displayed for monitoring by the ACU.

For testing the devices, a series of 8 healthy school children was tested. For determination of oxygen consumption, the expired air was collected during the last minute of exercise by the Douglas bag technique, O_2 was analyzed with Rapox and CO_2 with Capnograph (Godart-Statham). The plasma lactic acid was determined by the Boeringer UV test at maximal exercise and 4 min after exercise.

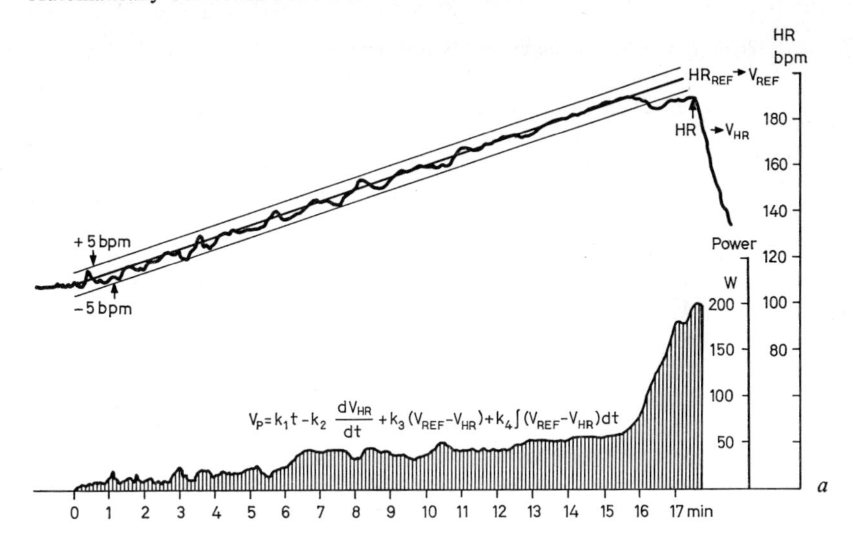

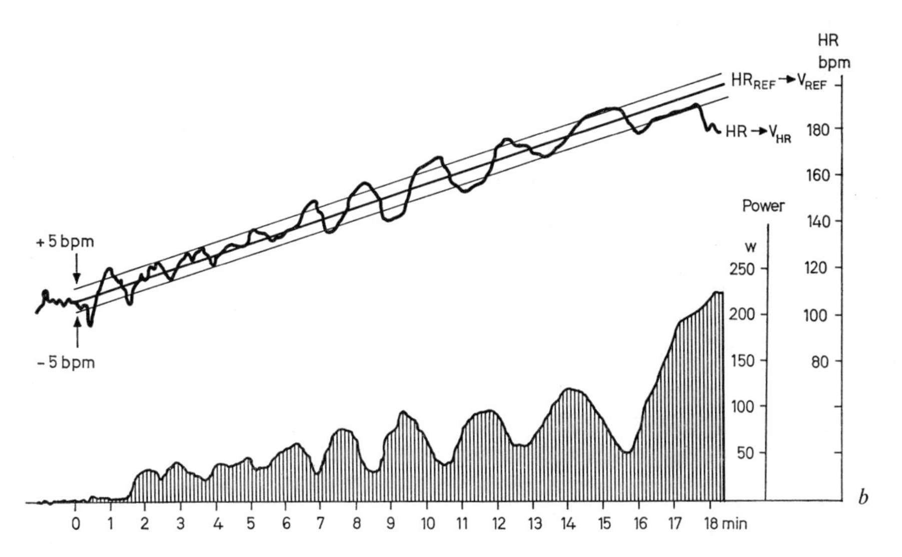

Fig. 2. Graphic output of the PCXT depicting the instantaneous heart rate and braking power. Examples of a proper (*a*) and oscillating (*b*) automatic control. The control equation is indicated in *a*.

Table I. Results of the automatically controlled PCXT in a preliminary series of 8 children

Subject	Sex	Age years	Weight kg	BSA m^2	TW Wmin	WL_{last} W	HR_{last} bpm	$V_{O_2}last$ l/min	$B\text{-}Lact_e$ mq %	$B\text{-}Lact_{e+4}$ mq %
1	M	11	35.8	1.2	1,520	170	190	2.45	38	29
2	M	12	40.5	1.4	450[1]	200	180	3.12	58	41
3	M	13	40.0	1.3	1,900	200	190	1.46	44	30
4	M	13	43.7	1.4	2,400	240	185	3.23	41	47
5	F	12	35.0	1.2	440[1]	110	185	1.93	14	9
6	F	12	56.5	1.6	1,350	215	190	3.04	49	53
7	F	13	56.0	1.6	1,000[1]	240	190	3.76	34	34
8	F	14	50.0	1.5	1,720	240	190	1.70	55	55

[1] Continued in manual control.

Experience of the Test Run

Among these very first test cases, there were 5 children in which the set-up functioned in the predicted way. A smooth rise of the heart rate and braking power was automatically generated (fig. 2A).

The results of the whole series are presented in table I. Initially, in 3 cases a proper way of using the automatic control was not found. A progressive oscillation of the system was generated (fig. 2B). The test subjects felt this very uncomfortable and the test was continued in manual (semiautomatic) control.

The first impression based on this pilot study was that the PCXT can be fully automated. However, the optimal combination of the terms in the differential equation was not satisfactorily solved. Sinus arrhythmia typical of children with low heart rate [4], seems to disturb the automatic regulation. Therefore, testing of subjects with very unstable heart rate presumes prominent k_4, i.e. strong filtering of heart rate at the beginning of the test is required.

The values of maximum heart rate indicate that, as it has also been stated previously, in children the use of working capacity corresponding to the rate of 170/min (W_{170}) is definitely measurement at submaximal level [3]. On the basis of the measured values of the maximal oxygen uptake and plasma lactic it appeared that the level of maximal exertion could be reached in most of our cases. MOCELLIN *et al.* [3] have shown that the oxygen uptake is lower in a continuous increase of load than in a stepwise ergometric test. This phenomenon might also influence our results. In further investigations, we have found that the availability of this automatic testing system greatly enchances the practical work required in studying the potential value of the PCXT in children.

References

1 Arstila, M.: Pulse-conducted triangular exercise-ECG test. A feedback system regulating work during exercise. Acta med. scand. *529:* suppl., p. 191 (1972).

2 Arstila, M.; Viherä, P., and Välimäki, I.: Semiautomatic controller of linear heartrate acceleration in ergometric exercise test. Proc. 1st natn. Meet. on Biophys. Biotech., Helsinki 1973, p. 159.

3 Mocellin, R.; Lindemann, H.; Rutenfranz, J., and Sbresny, W.: Determination of W_{170} and maximal oxygen uptake in children by different methods. Acta paediat. Scand. *217:* suppl., p. 13 (1971).

4 Petäjoki, M.-L.; Arstila, M., and Välimäki, L.: Pulse-conducted exercise test in children; in Borms and Hebbelinck Children and exercise. Acta paediat. belg. *1974:* 40.

Dr. I. Välimäki, Cardiorespiratory Research Unit, University of Turku, *20520 Turku 52* (Finland)

Medicine Sport, vol. 11, pp. 52–55 (Karger, Basel 1978)

The Rate of Growth in Maximal Aerobic Power of Children in Norway

K. L. ANDERSEN, J. RUTENFRANZ and V. SELIGER

Laboratorium fir Miljøfysiologi, Oslo

Physical performance capacity was defined by the WHO Expert Committee [1968] and was considered an important fitness variable. The Committee suggested the use of maximal aerobic power and underlying functions to describe health from a positive point of view.

In order to bring about information of population parameters with regard to achievement in exercise fitness during the period when children are attending schools, a study was initiated in Norway in 1969 on a cross-sectional and longitudinal basis. In this paper, results from a longitudinal study of children's fitness are presented.

Methods

The population of Lom community was selected for this study, because Lom represents a typical Norwegian rural inland district, and the people of Lom form a rather stable population.

The initial examination was undertaken in early fall 1969 in the sample of 8-year-old children. In the four subsequent years, tests were performed once a year at the same time of year. The physical characteristics are given in table I.

Maximal oxygen uptake was determined by exposing the subjects to increasing work loads. A bicycle ergometer of the mechanical braking type was used. Two submaximal and one maximal or hypermaximal load were performed. Work at submaximal loads lasted for 5 min, measurements were taken during the last minute of this period. The maximal loading lasted for 3 min and the measurements were taken during the last minute. The expired air was collected into Douglas rubber bags. The air volume was measured by using a wet gas meter. Aliquot samples of expired air was analyzed for CO_2 and O_2 by the Scholander gas analyzer or by infrared and paramagnetic principle, checked by the Scholander method.

Table I. Growth in basic anthropometric indices and in maximal aerobic power

Age, years		N	Height, cm		Weight, kg		Body fat, %		Maximal oxygen uptake					
									l/min		ml/min · kg		ml/min · kg LBM	
x̄	SD		x̄	SD	x̄	SD	x̄	SD	x̄	SD	x̄	SD	x̄	SD
Boys														
8.4	0.3	29	130.8	4.5	27.4	3.9	19.2	3.6	1.44	0.19	52.7	3.9	65.5	4.5
9.4	0.3	29	136.6	4.8	31.0	4.5	18.9	3.1	1.59	0.24	51.4	5.2	63.8	6.3
10.4	0.3	31	141.2	5.0	33.9	5.0	20.1	4.2	2.02	0.30	60.0	6.5	75.5	7.6
11.4	0.3	28	145.8	5.7	36.8	6.6	19.3	4.4	2.07	0.31	56.9	6.1	70.6	6.3
12.3	0.3	30	150.4	6.0	40.3	6.9	19.4	3.9	2.31	0.34	58.0	8.0	72.0	8.6
Girls														
8.2	0.3	33	130.1	4.5	26.7	3.5	23.3	2.9	1.25	0.20	47.4	7.0	61.3	8.3
9.3	0.3	33	136.0	5.1	30.8	4.1	23.0	3.0	1.48	0.19	48.5	6.6	63.2	8.5
10.3	0.3	34	141.3	5.5	34.4	4.6	24.3	3.4	1.19	0.23	52.4	6.6	69.7	8.2
11.2	0.3	34	145.5	10.3	38.1	6.4	24.4	3.5	1.88	0.22	50.1	5.9	66.4	6.2
12.2	0.3	34	152.5	6.5	42.6	7.2	24.1	3.5	2.26	0.32	53.6	6.8	71.0	8.0

Results

The relationship between oxygen uptake and work rate at submaximal loadings was estimated by calculations of correlation coefficient and variables of linear regressions separately for boys and girls and for each year of testing. The oxygen uptake at 300 kpm/min is 0.9–1.0 l/min, at 600 kpm/min 1.45–1.55 l/min and at 900 kpm/min 1.95–2.15 l/min (table I). These values are in agreement with the results relating oxygen uptake to work loads previously observed in the cross-sectional part of this investigation. The values compare well with other published data on the subject [ÅSTRAND, 1956].

Growth curves are fitted by using the calculated means at each age and drawn separately for boys and girls in absolute terms (l/min) as well as expressed in relation to body weight and lean body mass. By assuming a linear relationship between $\dot{V}O_2$ max. and age for subjects 8–12 years, linear regression equations are estimated.

The rate of growth changes with age is of an irregular manner. On the average for the 4-year interval during which measurements were taken annually, the boys gained 0.23 l/min per year against 0.25 l/min per year for the girls.

The children who were tested on a longitudinal basis reached fitness values considerably higher than expected from the cross-sectional data. The differences are statistically significant at ages above 10 years for the boys as well as for the girls.

Discussion

By assuming a linear relationship between $\dot{V}O_2$ max. and age during growth in children from 8 to 12 years, a line connecting mean values at age 8 and 12 is drawn, and corresponding values for the ages in between are named 'expected' values. The positive and negative deviations from the expected values occurs both for girls and boys. The oscillations seem to be quite irregular, and range from −4.7 to +8.4% for boys, and from −6.0 to +2.3% for the girls. A similar age-dependent variation in growth of exercise fitness is reported by MOCELLIN *et al.* [1971]. These annual oscillations in maximal aerobic power must be explained in terms of variations in environmental stimulations including behaviour and variations in habitual physical activity.

Relative values of maximal oxygen uptake in the examined population sample of children are higher than values presented by SELIGER *et al.* [1973]. They are very similar to values presented by ÅSTRAND [1956] and those found by RUTENFRANZ [1974]; a high percentage of very physically active children in these two examined groups is very likely.

The longitudinal measurements showed that development in maximal aerobic power outbalanced the growth in body size, so that the fitness expressed on the basis of body weight as well as on lean body mass increased as the children became older.

The improved fitness with time can only be explained by considering changes in environmental stimulation. These changes take place in a direction which promotes physiological adaptation, and only small contributions come from morphological adjustment in terms of indices of body size. Obviously, qualitative changes also take place in the active cell mass in adaptive responses to environmental stimulations.

References

Åstrand, P. O.: Human physical fitness with special reference to sex and age. Physiol. Rev. *36:* 307–312 (1956).

Mocellin, R.; Rutenfranz, J. und Singer, R.: Zur Frage von Normwerten der körperlichen Leistungsfähigkeit (170) im Kindes- und Jugendalter. Z. Kinderheilk. *110:* 140–165 (1971).

Rutenfranz, J.: Unpublished data (1974).

Seliger, V., *et al.*: Physical fitness of the Czechoslovak 12- and 15-year-old population. Acta paediat., Stockh. *217:* suppl., pp. 37–41 (1971).

WHO Export Committee: Exercise tests in relation to cardiovascular function. Report of a WHO meeting. Tech. Rep. Ser., Wld Hlth Org. 388 (1968).

Dr. K.L. Andersen, Laboratorium fir Miljøfysiologi, Eckersbergsgt 30–32,1, *Oslo 2* (Norway)

Medicine Sport, vol. 11, pp. 56–64 (Karger, Basel 1978)

A Children's Test of Fitness

D. A. BAILEY and R. L. MIRWALD

College of Physical Education, University of Saskatchewan, Saskatoon, Sask.

Introduction

The Canadian Home Fitness Test (CHFT) has now been made available to the Canadian public through bookstores and other agencies, and negotiations for distribution internationally are presently in the initial stages. The test is a motivational and educational device designed to increase personal physical activity. The CHFT is a self-administered test protocol in which the participant steps at an age- and sex-specific rhythm controlled by recorded music, then palpates the pulse immediately following activity. The development of the test and validation procedures have been well documented [BAILEY *et al.,* 1976; SHEPHARD *et al.,* 1976] and a procedure for the translation of the CHFT assessment into a program of graded exercise suited to the individual has been developed [JETTE, 1975].

Validation of the procedure has shown a correlation of 0.72 with the results of a standard bicycle ergometer test [BAILEY *et al.,* 1976]. JETTE *et al.* [1976] have also studied the CHFT using a multiple regression analysis and report that maximal oxygen intake in a defined population can be predicted with a multiple r of 0.905 using the energy cost of the attained stepping rate and age. The test has norms established for the age range 15–69 years.

The following report presents the findings of a project to extend the concept of the CHFT down to the junior high school age range of 11–14 years.

The Sample

A total of 212 subjects including 98 females and 114 male participants ranging in age from 11 to 14 years made up the sample. The children were drawn from six separate elementary schools, from the Riversdale Track Club, and from the Lions Speed Skating Club.

Exercise Test Procedure

The test utilized a double 8-inch step, similar to step heights found in the average Canadian home. Cadences for stepping rates were designed to elicit a post-exercise heart frequency corresponding to approximately 60% of the predicted maximum aerobic power for boys, and 65% for girls (see Appendix for details).

Subjects were tested in groups of four using a bank of four double 8-inch steps. Each subject was coupled to a four-channel electrocardiograph/cardiotachometer unit. The instantaneous pulse rate and ECG waveform were monitored continuously. Instantaneous heart rates were digitally displayed by cardiotachometer (precision $\pm 1\%$ to heart rates of 150/min, $\pm 2\%$ at higher heart rates). The accuracy of the individual cardiotachometer units were checked daily, using a pulse wave simulator, and at no stage were discrepancies greater than 1 beat/min noted.

The subjects began the step test by facing the step with both feet on the floor. When the musical tape with a rhythm set at 114 beats/min was begun, the subject stepped up onto the first 8-inch high step with one foot on the first beat, onto the top step with the opposite foot on the second beat, and brought both feet together on the top step on the third beat (standing erect, legs straight). The subject stepped down to the first step with one foot on the fourth beat, stepped down to the floor with the opposite foot on the fifth beat and brought both feet together on the floor on the sixth beat. This sequence represented one cycle of stepping, a pattern which was followed for an initial 6-minute period of exercise (fig. 1).

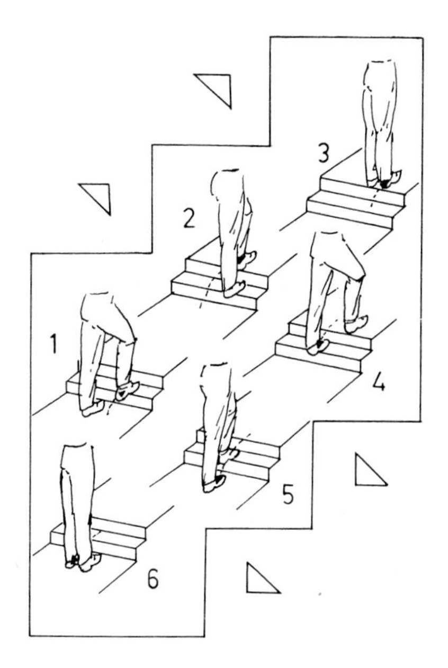

Fig. 1. Stepping sequence performed on double 8-inch steps to a six-count musical rhythm.

10 sec after the 6-min bout of exercise, a recorded voice signalled the commencement of pulse counting at the wrist or over the carotid artery. A second signal marked the end of the 10-sec counting period (20 sec post exercise). 10 sec later, those subjects who had not exceeded a heart rate of 162 commenced stepping for an additional 3 min at a faster cadence of 120 beats/min. The decision on whether subjects would proceed for an additional 3 min was based on the heart frequency obtained from the cardiotachometer using the average of 10 successive heart beats 'frozen' 15 sec after the cessation of exercise. The subject-obtained pulse count (taken from 10 to 20 sec following cessation of exercise) was subsequently compared to this accurate digitally recorded count for use in determining the pulse counting accuracy of the children.

Results

1. Safety. All children completed the initial 6-min portion of the test without incident. This held true for those children who were allowed to step for 9 min. Because of the submaximal nature of the work task, this test is clearly less strenuous than some daily activities that many of the children are involved in, and is safe for any normal child.

2. Practicality. Children in the age range tested had no difficulty in mastering the stepping rhythms. In general, the children seemed to enjoy taking the test.

3. Validity. Previous reports on a similar type of stepping test for adults [BAILEY *et al.*, 1976; SHEPHARD *et al.*, 1976] have demonstrated a good relationship between step and bicycle ergometer results. This relationship held reasonably stable over the age range 15–69.

To further check validity, the physical activity of the children was rated by their teacher according to the involvement of the children in sport activities. It can be seen from figures 2 and 3 that the test discriminated quite well between children classified according to physical activity patterns.

Figures 4 and 5 present the results of the test when analyzed according to school attended. The Director of Physical Education was asked to rate each school according to the quality of the physical education program within the school. Figure 4 shows that there is a high degree of relationship between test results for the boys and the rating of the Director of Physical Education. However, this relationship breaks down when the girls' results are analyzed, as demonstrated in figure 5. Why this should be is probably best left for others to explain.

4. Pulse-taking accuracy. A practical problem associated with a self-administered test is the inability of people to accurately take their own pulse. In establishing the Canadian Home Fitness Test the following facts were

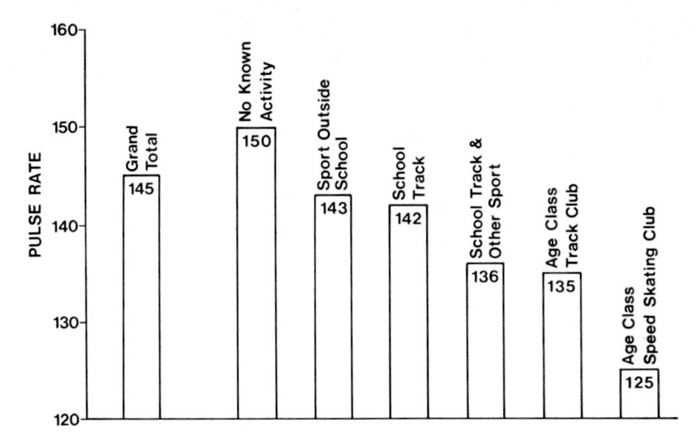

Fig. 2. 6-min recovery pulse for boys classified according to physical activities.

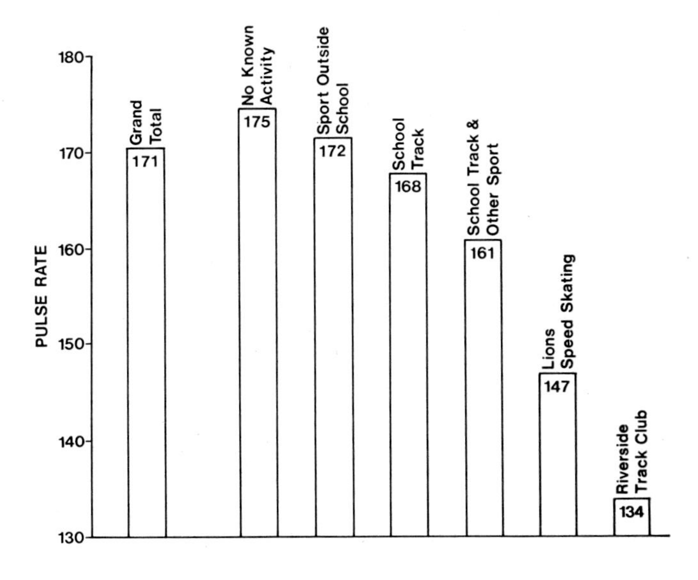

Fig. 3. 6-min recovery pulse for girls classified according to physical activity.

noted: (1) Regardless of the sex or age group, people who are untrained at pulse counting cannot take their own pulse accurately. However, it was demonstrated that accuracy improved considerably with practice. (2) The error in pulse counting was always in one direction with a consistent tendency toward underestimation.

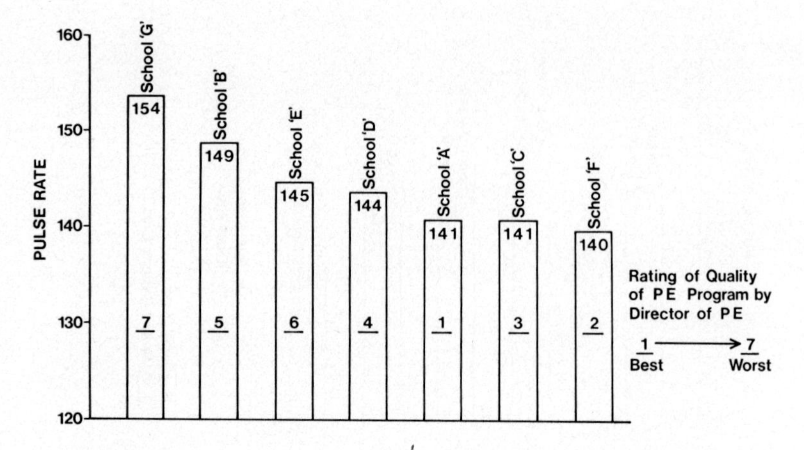

Fig. 4. 6-min recovery pulse for boys classified according to school attended.

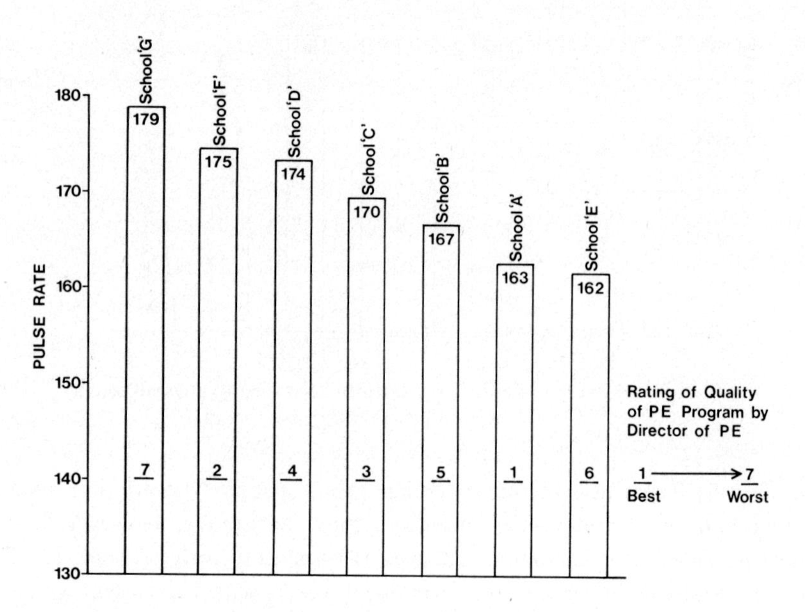

Fig. 5. 6-min recovery pulse for girls classified according to school attended.

Table I. Comparison of mean recovery heart rate/min for three groups: actual[1] vs manual[2]

	Group	N	Actual[1]	Manual[2]	Mean diff.	t-value	Corre-lation
Experienced pulse counters	I	18	134.7	127.3	7.3	1.21	0.94
Partner pulse counters	II	80	153.1	142.1	11.0	3.26*	0.76
Self-counters	III	114	151.9	138.3	13.6	7.75*	0.37

* Significant at 0.01 level.
[1] Recovery HR/min taken from cardiotachometer 'frozen' 15 sec post exercise.
[2] Recovery HR/min taken manually from 10 to 20 sec post exercice.

To investigate pulse-taking accuracy in children, the design of the present investigation provided for the division of the total sample into three groups: (a) Experienced pulse counters. Children in this group were well trained in pulse counting. Most of them regularly used pulse counting to monitor their training programs for track or speed skating. (b) Partner pulse counters. In this group, a youngster was selected to count the pulse of children taking the test. That is, the pulse count was taken by another child. (c) Self-counters. In this group, the youngsters taking the test counted their own pulse.

Table I presents the results of the pulse-counting investigation. Essentially, it reaffirms the results of the previous Canadian Home Fitness Test Study that the difference between cardiotachometer and manual heart rates is in one direction and with practice the accuracy improved. In addition, the results indicate that: (1) Experienced pulse counters are able to take their pulse accurately (within one beat for the 10-sec period). The t-value for this group is not significant, indicating no statistical difference between the cardiotachometer heart rate and the manual heart. (2) Partner pulse counters are an improvement over inexperienced self-counters.

On the basis of this subinvestigation of pulse counting, the following conclusions would appear to be warranted: (1) Children in this age range with experience in taking their own pulse can get an accurate assessment of heart rate. Therefore, prior to administration of the test children should be trained and given experience in pulse counting. (2) If inexperienced children are to be tested, an experienced partner should be used as a pulse counter.

Table II. Stepping time of children categorized according to physical activity

Subjects	Stepping time 6 min	Stepping time 9 min	Percent stepping 9 min
Boys			
No known activity	50	40	80.0
One activity outside school	22	19	86.4
School track	14	14	100.0
School track plus an outside activity	16	16	100.0
Riversdale Track Club	6	6	100.0
Lions Speed Skating Club	6	6	100.0
Total group	114	101	88.6
Girls			
No known activity	54	10	18.5
One activity outside school	12	3	25.0
School track	5	3	60.0
School track plus outside activity	18	12	66.6
Riversdale Track Club	5	5	100.0
Lions Speed Skating Club	4	4	100.0
Total group	98	37	37.8

Table III. 10-sec recovery heart rate norms for a simple children's test of physical fitness. Heart rate by palpation from 10 to 20 sec following 6 min of stepping at a cadence of 114/min

Boys		Girls
28 or higher →	needs activity	← 29 or higher
27, 26, 25 →	can improve	← 28, 27, 26
24, 23 →	good shape	← 25, 24
22 or lower →	super fit	← 23 or lower

5. *Duration of test.* Some children had recovery heart rates ≤162 for the 10-second period from 10 to 20 sec after the cessation of 6 min of stepping. These children were allowed to continue to step for an additional 3 min at a slightly faster cadence. Of 114 boys, 101 (88.6%) stepped for a total of 9 min. Less than half of the girls continued on to 9 min, 37 out of 98 girls (37.8%).

Table IV. 10-sec recovery heart rate norms for a simple children's test of physical fitness. Heart rate by palpation from 10 to 20 sec following 6 and 9 min of stepping at a cadence of 114 and 120/min (girls' values in parentheses)

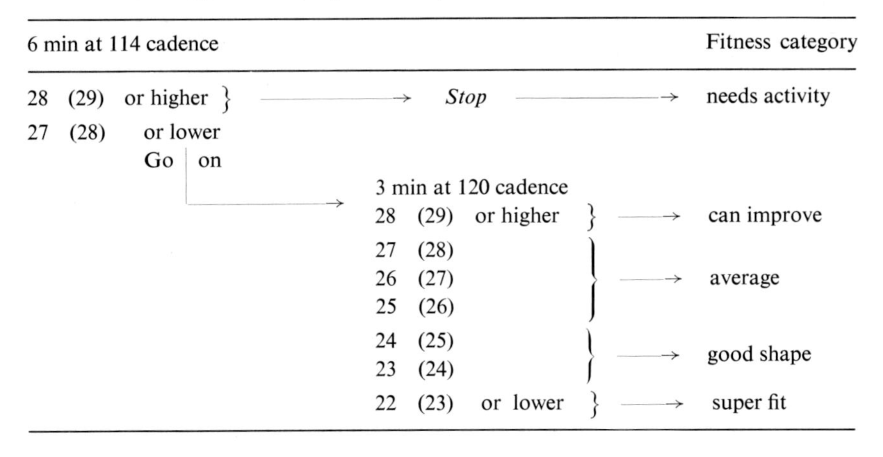

6 min at 114 cadence	3 min at 120 cadence	Fitness category
28 (29) or higher →	Stop →	needs activity
27 (28) or lower Go on →	28 (29) or higher →	can improve
	27 (28) 26 (27) 25 (26) →	average
	24 (25) 23 (24) →	good shape
	22 (23) or lower →	super fit

Table V. Recovery pulse values (mean ± 1 SD) for children categorized according to physical activity pulse values in beats/min by cardiotachometer 'Frozen' at 15 sec post exercise

	Boys 6-min values		Girls 6-min values		Boys 9-min values		Girls 9-min values	
	n	M SD	n	M SD	n	M SD	n	M SD
No known activity	50	150 ± 16	54	175 ± 14	40	146 ± 11	10	156 ± 7
One activity outside school	22	143 ± 13	12	172 ± 16	19	145 ± 10	3	153 ± 12
School track	14	142 ± 15	5	168 ± 16	14	139 ± 11	3	158 ± 6
School track plus an outside activity	16	136 ± 9	18	161 ± 14	16	136 ± 8	12	151 ± 10
Riversdale Track Club	6	135 ± 6	5	134 ± 6	6	136 ± 8	5	139 ± 9
Lions Speed Skating Club	6	125 ± 6	4	147 ± 6	6	127 ± 8	4	155 ± 15
Total group	114	145 ± 15	98	171 ± 15	101	143 ± 11	37	154 ± 9

Table II shows the percentage of children in each activity group who were allowed to step for 9 min. Using the same heart rate cut off of 162 for both boys and girls clearly discriminated against the girls as is evident from table II. If the test is to be used as a 9-min test, then the heart rate cut off for girls should be raised to 168 (28 for the 10-sec count).

6. Determination of norms. The test can be administered as a simple 6-min test. On this basis, the general categories as listed in table III may be appropriate. If the test is to be just 6 min in duration, a longer pulse count could be used. If the 9-min test is used, the categories outlined in table IV are possibilities. The values needed to construct norm tables are contained in table V.

References

BAILEY, D.A.; SHEPHARD, R.J., and MIRWALD, R.L.: Validation of a self-administered home test of cardio-respiratory fitness. Can. J. appl. Sports Sci. *1:* 67 (1976).
JETTE, M.: An exercise prescription program for use in conjunction with the Canadian Home Fitness Test. Can. J. publ. Hlth *66:* 461 (1975).
JETTE, M.; CAMPBELL, J.; MONGEON, J., and ROUTHIER, R.: The Canadian Home Fitness Test as a predictor of aerobic capacity. Can. med. Ass. J. *114:* 680 (1976).
SHEPHARD, R.J.: World standards of cardio-respiratory performance. Archs envir. Hlth *13:* 664 (1966).
SHEPHARD, R.J.; BAILEY, D.A., and MIRWALD, R.L.: Development of the Canadian Home Fitness Test. Can. med. Ass. J. *114:* 675 (1976).

Appendix A

Basis of a Step Test for Children (Grades 6–8)

The average fitness is based on rounded data from SHEPARD [1966]. The gross aerobic power (ml/kg/min) is further reduced by 3 ml/kg/min to yield an approximate net aerobic power; the appropriate stress for the test is 60% of the gross aerobic power less 3 ml/kg/min for boys and 65% for girls. This is necessary to have the same cadence for both sexes.

Assume a weight of 45 kg at age 13, a mechanical efficiency of stepping of 14% at this age, and a caloric yield of 5 kCal/l of oxygen. On this basis, the load needed to produce 60% VO_2 max. for boys and 65% VO_2 max. for girls is 19 ascents/min on a double 8-inch step.

Dr. D.A. BAILEY, College of Physical Education, University of Saskatchewan, *Saskatoon, Sask.* (Canada)

Medicine Sport, vol. 11, pp. 65–71 (Karger, Basel 1978)

Muscle Metabolic Studies of a Girl with McArdle-Like Symptoms

Bengt O. Eriksson, Olle Hansson, Jan Karlsson and Karin Piehl

Department of Pediatrics, University of Göteborg, Göteborg, and
Department of Physiology, Gymnastik och Idrottshögskolan, Stockholm

In 1951, the English neurologist McArdle described a syndrome characterized by myopathy and muscular fatigue and pain during exercise. Further exercise produced muscle spasms and swelling [McArdle, 1951]. This inability to continue exercising was accompanied by the absence of any increase in the level of blood lactate. In 1959, this inability to increase the concentration of blood lactate was shown to be due to an inability to break down glycogen into lactate in muscle tissue because the enzyme phosphorylase was lacking [Mommaerts *et al.*, 1959; Schmidt and Mahler, 1958]. This glycogenosis has subsequently been the subject of many studies providing information both on the abnormal metabolism in these cases and on normal muscle metabolism [Gruener *et al.*, 1968; Mellick *et al.*, 1962; Pearson *et al.*, 1961; Pernow *et al.*, 1967; Porte *et al.*, 1966; Rowland *et al.*, 1963; Schimrig *et al.*, 1967; Tobin and Coleman, 1965].

The muscle metabolic characteristics of this syndrome have recently been described by Jorfeldt *et al.* [unpublished] and Wahren *et al.* [1973]. Approximately 20 cases of the McArdle's syndrome have hitherto been reported. No case is presently known in Sweden.

This brief report deals with a girl displaying some of the features of the McArdle syndrome without actually suffering from true McArdle syndrome.

Material, Procedure and Methods

The girl, C.N., born in 1962, was the second of 3 children. Both parents were healthy; however, her mother had also displayed symptoms resembling the girl's. The mother's pregnancy was normal, but delivery was pre-term, and the infant weighed 1,580 g at birth.

Her 'catch up' in growth was normal. Thus, her weight and behaviour had normalized by the age of 1 year. Her medical history was uneventful up to the age of nine. It was then that her symptoms made their debut. These consisted of painful swelling of the muscles during exercise, most pronounced in the deltoid and triceps muscles. The symptoms gradually disappeared when she allowed the affected muscles to rest. However, the symptoms generally failed to abate fully until the next day. No change was noticed in the colour or smell of her urine.

Since the intensity of symptoms increased somewhat, the girl was admitted for a thorough examination. Physical findings were normal. She proved to be somewhat smaller than average (height 149 cm, weight 35.3 kg). A routine neurological examination was unable to disclose any abnormalities. Routine analyses of blood and urine, EEG, ECG and scull X-rays were normal. No myoglobin was found in the urine. Electromyograms from the most severely affected muscles (the deltoid, tibialis anterior, and interosseus dorsalis I) were normal. Conduction speeds in the ulnar and peroneal nerves amounted to 68 and 57 m/sec, respectively. Thus, no neurological grounds for her symptoms were found.

Since the symptoms resembled McArdle's syndrome, some tests were made of her ability to form lactate. She performed a total of 4 different exercise tests. Three of the tests were performed during her first admission in 1973 and the fourth in 1974. Each test featured a gradually increasing work load, the loads being 150, 300 and 450 kpm/min. Exercise at each load was carried out for 6 min. Her heart rate was checked every second minute and blood samples were taken from a pre-warmed fingertip at the end of each work load. The blood lactate concentration was determined using an enzymatic method [Scholz *et al.*, 1959].

A further exercise test was performed at her second admission. This time, the girl exercised at a heavier load, 550 kpm/min. Muscle biopsies were taken at rest on this occasion, immediately after 450 and after 550 kpm/min. The samples were taken using a needle according to Bergström [1962] and immediately frozen in liquid nitrogen. Analyses for glycogen, glucose, glucose-6-phosphate, lactate, creatine phosphate (CP) and adenosine triphosphate (ATP) were made using a modified Lowry technique [Karlsson, 1971].

Muscle enzymes such as phosphorylase (Plas), phosphofructokinase (PFK), lactate dehydrokinase (LDH), and hexokinase (HK), were determined as previously described [Gollnick *et al.*, 1974]. In addition, muscle fibre composition was determined according to Gollnick *et al.* [1972].

Glycogen synthetase activity was determined according to a modified version of the method described by Thomas *et al.* [1968]; for details, see Adolfsson [1972].

Results

The 3 exercise tests in 1973 produced very similar heart rate responses (fig. 1). On the other hand, blood lactate concentration differed. The blood lactate level in the first test remained almost constant, despite the increasing load. This was also the case in the third test. The second test produced an increase up to 5.8 mmol/l. This test also resulted in a slightly slower heart rate response.

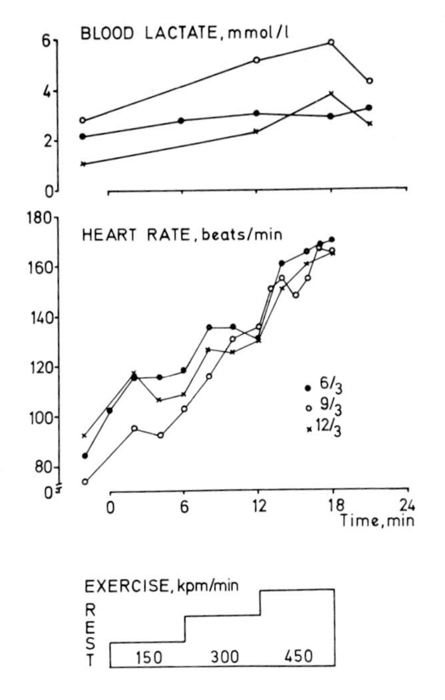

Fig. 1. Blood lactate concentration and heart rate at rest and at three work loads on three different days in exercise in 1973 by a girl with McArdle-like symptoms.

In the 1974 study (fig. 2), her blood lactate concentration was elevated at rest, i.e. 5.2 mmol/l, but decreased to 2.2 mmol/l when she started exercising. Further exercise produced an increase up to 4.9 mmol/l.

Muscle lactate concentration amounted to 4.6 mmol/kg w.w. at rest and increased to 5.9 and 6.1 mmol/kg w.w. at 450 and 550 kpm/min, respectively (fig. 2). The glycogen concentration decreased from 103 to 82 and 53 mmol/kg glucose units w.w. at 450 and 550 kpm/min, respectively. The glucose-6-phosphate concentration remained fairly constant at a low level throughout the exercise test. However, a slight increase was obtained at the highest work load. Glucose concentration, on the other hand, increased strikingly from 0.39 to 1.40 mmol/kg w.w. The concentrations of ATP and CP were somewhat high at rest, i.e. 6.4 and 20.8 mmol/kg w.w. ATP was unchanged by exercise but CP levels dropped to 9.6 mmol/kg w.w.

The activities of some of the glucolytic enzymes are given in table I. Glycogen synthetase activity (I + D) amounted to 2.3 mmol × kg^{-1} at rest,

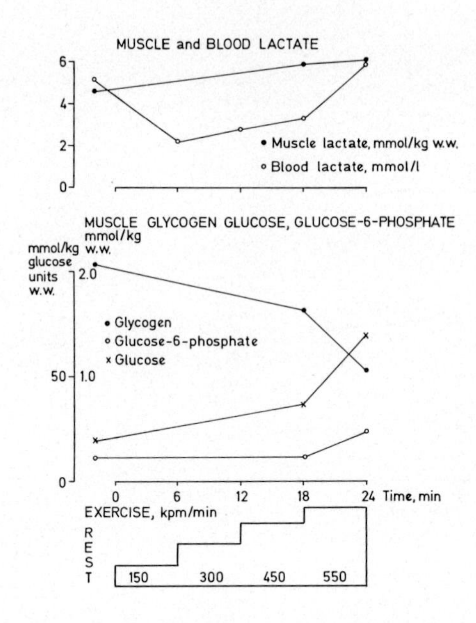

Fig. 2. Muscle and blood lactate concentrations and glycogen, glucose-6-phosphate and glucose concentrations in muscle tissue at rest and during exercise in a girl with McArdle-like symptoms.

Table I. Some values for muscle enzyme activity at rest and during exercise in a girl with McArdle-like symptoms. Normal values at rest for a young adult man are also included [GOLLNICK *et al.* 1974]

	Rest	450 kpm/min	550 kpm/min	Normal values
LDH (pyruvate lactate), mol/g × min w.w.	0.82×10^{-4}	0.78×10^{-4}	0.80×10^{-4}	$(0.83–1.18) \times 10^{-4}$
LDH (lactate pyruvate), mol/g × min w.w.	0.63×10^{-4}	0.48×10^{-4}	0.52×10^{-4}	$(0.19–0.27) \times 10^{-4}$
Hexokinase, mol/g × min w.w.	0.13×10^{-5}	0.11×10^{-5}	0.13×10^{-5}	$(0.19–0.25) \times 10^{-5}$
PFK, mol/g × min w.w.	0.15×10^{-5}	0.15×10^{-5}	0.16×10^{-5}	$(0.19–0.27) \times 10^{-6}$
Phosphorylase (a + b), mol/g × min w.w.	0.75×10^{-6}	0.65×10^{-6}	0.88×10^{-6}	$(0.50–1.12) \times 10^{-6}$

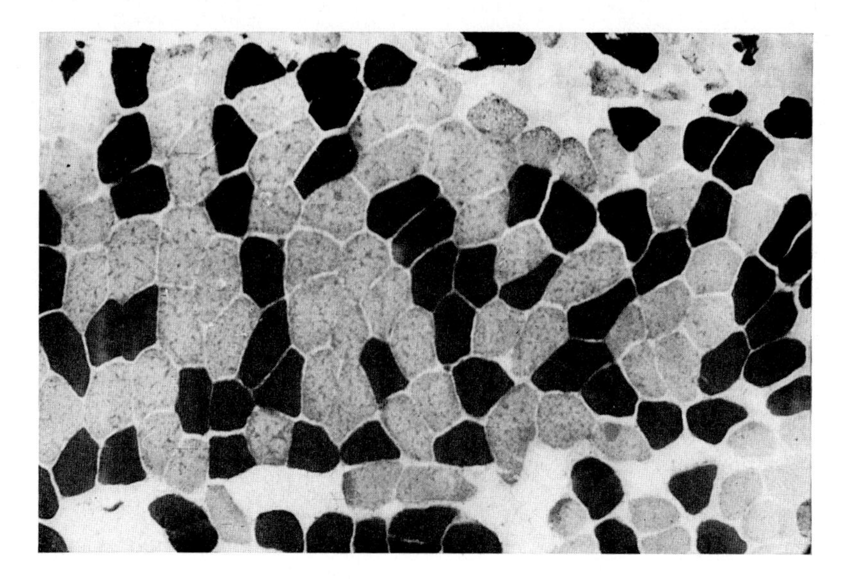

Fig. 3. Histochemical staining pattern for myofibrillar ATPase in the quadriceps muscle of a girl with McArdle-like symptoms. Lightly stained fibres are slow twitch fibres (ST fibres).

which is essentially within the normal range. The activity remained unchanged during work.

The histological determination of fibre types disclosed that 32% of the fibres were slow twitch (fig. 3).

Discussion

In many respects, the patient's clinical picture resembled the symptoms and signs ascribed to the McArdle syndrome. This syndrome is now well-known. Because of a lack of Plas, no glycogen is broken down in exercising muscles. Thus, in addition to ATP and CP, only available glucose and fatty acids can be used as 'fuel'. This results in a decreased work capacity and 'second wind phenomenon' as described by PERNOW *et al.* [1967].

Partial disturbances in glycolysis have also been described [ENGEL *et al.*, 1963; MELLICK *et al.*, 1962; PORTE *et al.*, 1966; TOBIN and COLEMAN, 1965]. These cases displayed a less pronounced increase in blood lactate concentration after ischaemic work. Many of these 'partial McArdles' were relatives of known cases of McArdle, and some of them later developed into typical cases of McArdle syndrome [JORFELDT *et al.*, unpublished].

The present patient is able to break down glycogen into lactate. However, very slight or no increase in blood lactate levels were found in some of the exercise tests. It is conceivable that this variation may be due to changes in phosphorylase activity from one day to another. However, this question cannot be answered as muscle specimens were obtained from only one of the four exercise tests.

High blood and muscle lactate values were obtained at rest in one of the exercise tests (fig. 2). The explanation for these values is obscure but it cannot be excluded that anxiety in relation to the tests was the major reason.

We do not know whether or not this patient will develop a classical McArdle in the future as has been described earlier [JORFELDT *et al.*, unpublished] or if she represents a new disease. Further studies are called for.

References

ADOLFSSON, S.: Regulation of glycogen synthesis in muscle; thesis, Göteborg (1972).

BERGSTRÖM, J.: Muscle electrolytes in man. Scand. J. clin. Lab. Invest. suppl. 68 (1962).

ENGEL, W.K.; EYERMANN, E.L., and WILHAMS, H.E.: Late onset of skeletal muscle phosphorylase deficiency: a new familiar variety with completely and partially affected subjects. New Engl. J. Med. *268:* 135 (1963).

GOLLNICK, P.D.; ARMSTRONG, R.B.; SAUBERT, C.W., IV; PIEHL, K., and SALTIN, B.: Enzyme activity and fiber composition in skeletal muscle of untrained and trained men. J. appl. Physiol. *33:* 312–319 (1972).

GOLLNICK, P.D.; SJÖDIN, B.; KARLSSON, J.; JANSSON, E., and SALTIN, B.: Human soleus muscle: a comparison of fiber composition and enzyme activities with other leg muscles. Pflügers Arch. *348:* 247–255 (1974).

GRUENER, R.; MCARDLE, B.; RYMAN, B.E., and WELLER, R.O.: Contractur of phosphorylase deficient muscle. J. Neurol. Neurosurg. Psychiat. *31:* 268–283 (1968).

JORFELDT, L.; PERNOW, B.; HAVEL, R.J.; SALTIN, B., and WAHREN, J.: Circulatory and metabolic changes during leg exercise in McArdle's syndrome (unpublished manuscript).

KARLSSON, J.: Lactate and phosphagen concentrations in working muscle of man. Acta physiol. scand. suppl. 358 (1971).

MCARDLE, B.: Myopathy due to a defect in muscle glycogen breakdown. Clin. Sci. *10:* 13–35 (1951).

MELLICK, R.S.; MAHLER, R.F., and HUGHES, B.P.: McArdle's syndrome: phosphorylase-deficient myopathy. Lancet *i:* 1045–1048 (1962).

MOMMAERTS, W.F.H.M.; ILLINGWORTH, B.; PEARSON, C.M.; GUILLORY, R.J., and SERAYDARIAN, K.: A functional disorder of muscle associated with the absence of phosphorylase. Proc. natn. Acad. Sci. USA *45:* 791–797 (1959).

Pearson, C. M.; Rimer, D. G., and Mommaerts, W. F. H. M.: A metabolic myopathy due to absence of muscle phosphorylase. Am. J. Med. *30:* 502–517 (1961).

Pernow, B.; Havel, R. J., and Jennings, D. B.: The second wind phenomenon in McArdle's syndrome. Acta med. scand. suppl. 472, pp. 294–307 (1967).

Porte, D.; Crawford, D.; Jennings, D. B.; Aber, C., and McIlroy, M.: Cardiovascular and metabolic responses to exercise in McArdle's syndrome. New Engl. J. Med. *275:* 406–412 (1966).

Rowland, L. P.; Fahn, S., and Schotland, D. L.: McArdle's disease. Hereditary myopathy due to absence of muscle phosphorylase. Archs Neurol., Chicago *9:* 325–342 (1963).

Schimrig, K.; Merteus, H. G.; Ricker, K.; Kühr, J.; Eyer, P. und Pette, D.: McArdle-Syndrom (Myopathie bei fehlender Muskelphosphorylase) Klin. Wschr. *45:* 1–17 (1967).

Schmid, R. and Mahler, R.: Chronic progressive myopathy with myoglobinuria: demonstration of a glucogenolytic defect in the muscle. J. clin. Invest. *38:* 2044–2058 (1958).

Scholz, R.; Schmitz, H.; Buecher, T. und Lampen, J. O.: Über die Wirkung von Nystatin auf Bäckerhefe. Biochem. Z. *331:* 71–86 (1959).

Thomas, J. A.; Schlender, K. K., and Larner, J.: A rapid filter paper assay for UDP-glucose-glycogen glycosyltransferase, including an improved biosynthesis of UDP-^{14}C-glucose. Analyt. Biochem. *25:* 486–499 (1968).

Tobin, R. B. and Coleman, W. A.: A family study of phosphorylase deficiency in muscle Ann. intern. Med. *62:* 313–327 (1965).

Wahren, J.; Felig, P.; Harel, R. J.; Jorfeldt, L.; Pernow, B., and Saltin, B.: Amino acid metabolism during exercise in McArdle's syndrome. New Engl. J. Med. *288:* 774–777 (1973).

Dr. B. O. Eriksson, Department of Pediatrics, Östra Sjukhuset, *S-416 85 Göteborg* (Sweden)

Medicine Sport, vol. 11, pp. 72–81 (Karger, Basel 1978)

Pulmonary Function and Spiroergometric Criteria in Scoliotic Patients before and after Harrington Rod Surgery and Physical Exercise

H. STOBOY

Orthopädische Klinik und Poliklinik der Freien Universität Berlin im Oskar-Helene-Heim and Institut für Leistungsmedizin, Berlin

Introduction

The most important consequence for patients with a major scoliotic chest deformity is cardiopulmonary insufficiency.

The first change in young patients is a diminished pulmonary function with a decreased vital capacity (57–90% of the predicted value) [MEISTER and HEINE, 1973; GUCKER, 1962; COLLINS and PONSETT, 1969; WESTGATE and MOE, 1969; SCHEIER, 1967]. Most of the authors agree that the degree of reduced vital capacity is strongly and negatively correlated with the angle of the lateral curvature (Cobb angle) [MEISTER and HEINE, 1973; MANKIN *et al.*, 1964; SCHEIER, 1967; WESTGATE and MOE, 1969; MEZNIK and KUMMER, 1970; HEINE and MEISTER, 1972; HILPERT and BILGE, 1973]. Maximum voluntary ventilation can be decreased to 55% [WESTGATE and MOE, 1969] or even more [SCHEIER, 1967] statistically depending on the Cobb angle [HILPERT and BILGE, 1973; WESTGATE and MOE, 1969].

According to MAKLEY *et al.* [1968], the decreased vital capacity and maximum voluntary ventilation are predominantly due to pulmonary restriction. According to BERGOFSKY *et al.* [1959], EULER [1951] and ULMER *et al.* [1970], alveolar hypoventilation leads to a constriction of alveolar vessels and to pulmonary hypertension. AEPPLI [1964] and CHAPMAN *et al.* [1939] state that in non-treated or insufficiently treated patients life expectancy is limited to 30–50 years [BOAS, 1923]. The aim of Harrington rod surgery and/or spinal fusion was to preserve respiratory function at the level of presurgery values [BERGOFSKY *et al.*, 1959; WESTGATE and MOE, 1969; MEISTER and HEINE, 1973]. Some authors report in spite of an extensive conservative or surgical treatment, a further decrease of vital capacity between 10 and 21% [GUCKER, 1962; WESTGATE and MOE, 1969].

Surgical treatment was combined with exercise therapy, to investigate the changes of spirographic and spiroergometric values during a rehabilitation period of 1.5 years.

Methods and Material

11 patients, aged 13–20 years, with idiopathic scoliosis in which the Cobb angle ranged between 37 and 130° ($\bar{x} = 65.5°$) underwent Harrington rod surgery and spinal fusion with bone graft. After surgical treatment, the mean angle was decreased to 39.8° (15–90°).

The usual spirographic investigations were carried out at the admission, immediately before surgery, 9 weeks later, 1 year after surgery when the extension cast was removed, and 6 months later. At the same time with the exception of presurgery investigation spiroergometric values were measured at ½, 1 W/kg and maximum workload levels.

The exercise program started immediately after the admission. Twice daily dynamic exercises, for instance rope skipping, were carried out. In 3 series (30–50 sec) the heart rate was raised between 150 and 160/min. During the intervals, the heart rate did not drop below 130/min. The exercises were also performed with a Cotrel extension cast and 6 days before surgery during complete bedrest. During the 1 week presurgical bedrest exercises were done with small one-hand barbell weights and coordinated movements of the legs. Seven days after surgery, exercising started again and was continued approximately 4 weeks later after the application of a Cotrel extension cast. Eight weeks after surgery, the patients had to cycle an ergometer 6 times a week for a 10-min period. For 3 months, the workload was approximately 1.0 W/kg, and in the rest of the 1-year period 1.5–2 W/kg. One year after surgery, the cast was removed. The patients were informed that dynamic exercises and ergometer cycling had to be continued. Additionally, they were ordered to swim a half-hour twice per week.

Results

Vital capacity (VC) and forced vital capacity (FVC) did not change significantly (fig. 1).

Maximum voluntary ventilation (MVV) increased with a highly significant difference ($p < 0.025$) between admission and end of the observation period (fig. 2).

In the first investigation, heart rate at rest was far above the normal level (110/min) and increased in the post-surgical period (118/min) significantly ($p < 0.025$). During the following period, it approached normal range. Between the first and the last measurement, it was markedly diminished

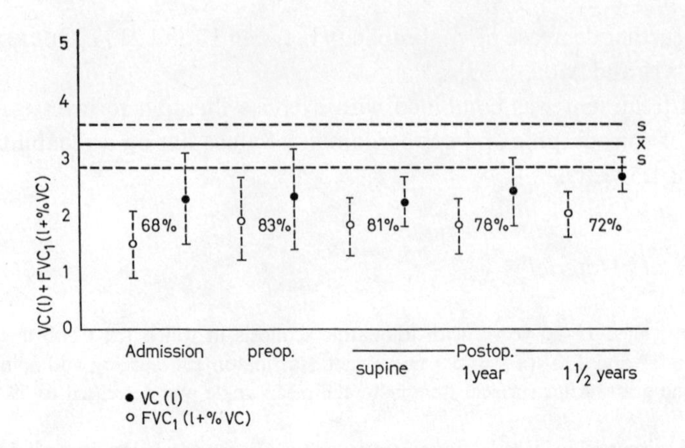

Fig. 1. Changes of vital capacity (VC) and forced vital capacity (FVC_1) at admission and during the observation period. $\bar{x}$ = Mean values of normals; s = standard deviation of normals.

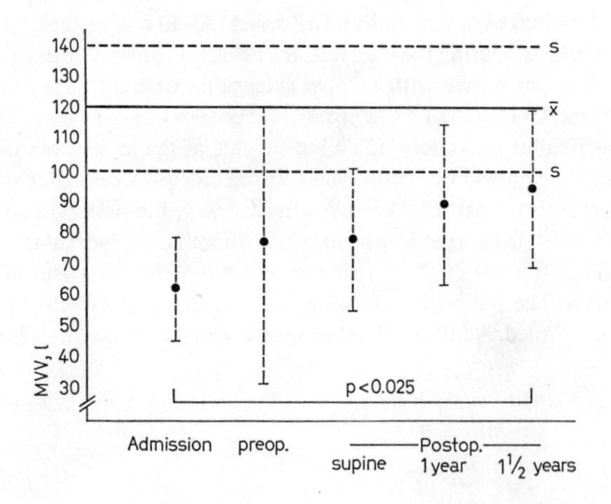

Fig. 2. Increase of maximum voluntary ventilation (MVV) during the observation period. $\bar{x}$ = Mean values of normals; s = standard deviation of normals.

(103/min, $p < 0.05$). During the whole observation period, maximum heart rate corresponded to the mean value of normal subjects. Therefore, the difference between heart rate at rest and maximum heart rate in the last investigation was, at 85.5%, the largest (fig. 3).

The minute volume of respiration ($\dot{V}$) during maximum workload increased during rehabilitation from 58 to 68 l/min without any statistical signifi-

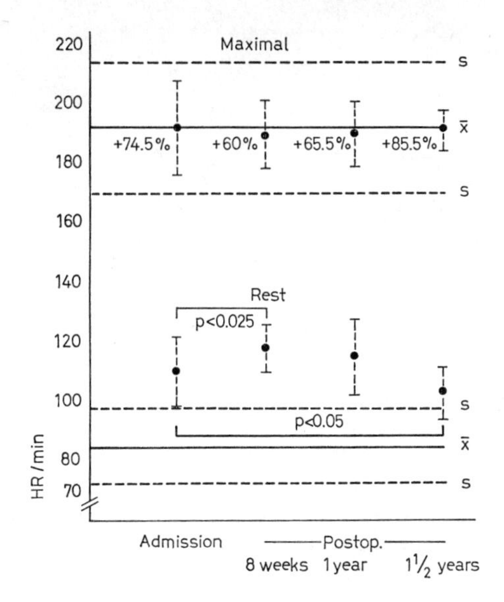

Fig. 3. Maximal (above) and rest heart rate (below) during the rehabilitation period. $\bar{x}$ = Mean values of normals; s = standard deviation of normals.

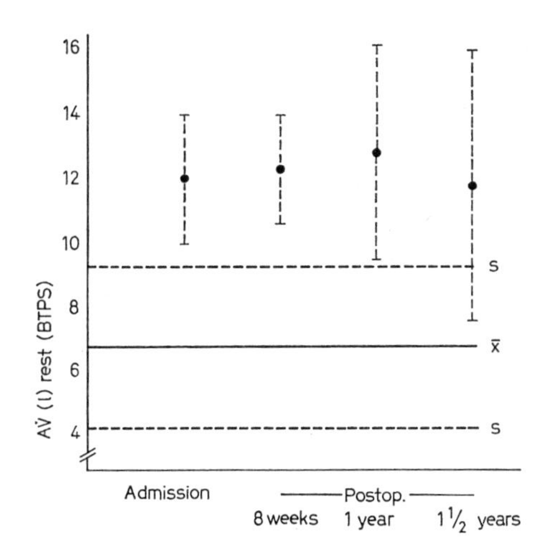

Fig. 4. During the observation period no significant changes of minute volume of respiration could be observed. $\bar{x}$ = Mean values of normals; s = standard deviation of normals.

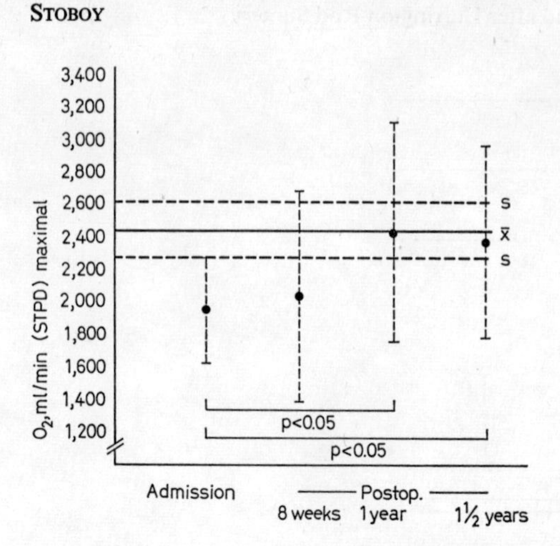

Fig. 5. Significant increase of the maximum O_2 uptake during the observation period. $\bar{x}$ = Mean values of normals; s = standard deviation of normals.

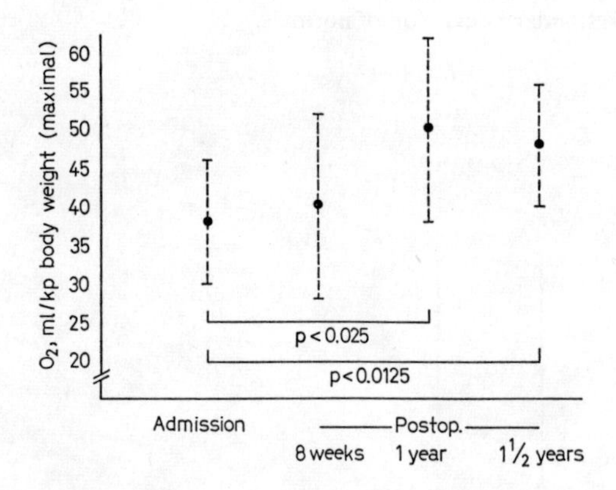

Fig. 6. Significant increase of maximum O_2 uptake related to body weight during the rehabilitation period.

cance because of the large standard deviation. These values were roughly 10% larger than normal. During rest, hyperventilation could be observed (fig. 4).

The mean maximum O_2 uptake was very poor at admission (1.9 l/min). One year and 1.5 year after surgery, it had increased significantly and was included in the normal range (2.4 l/min; $p < 0.05$) (fig. 5). This improvement

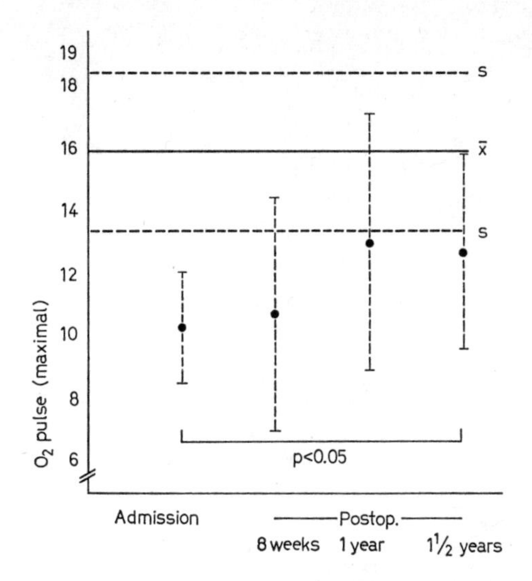

Fig. 7. Increase of the primarily diminished maximum O_2 pulse during the period of observation. $\bar{x}$ = Mean values of normals; s = standard deviation of normals.

of O_2 uptake is more pronounced when related to body weight ($p < 0.025$ or < 0.0125, respectively) (fig. 6).

Due to decreasing heart rate at rest and especially to the enhanced maximum O_2 uptake at the end of the observation period, the maximum O_2 pulse increased markedly from a mean value of 10.3–13.0 ($p < 0.05$), approaching the normal range (fig. 7).

The respiratory equivalent was at rest approximately 39, reached its lowest value with 26.5 at a workload of 1 W/kg, increasing during maximum performance to 28. During the whole course of rehabilitation, there were only small differences and no significant changes could be detected.

The respiratory frequency during maximum performance averaged 46/min and varied also only slightly (44–50/min).

Only the vital capacity, measured in absolute values or as percentage of predicted values, was at admission and at the end of the observation period highly negatively correlated to the Cobb angle ($r = -0.86$ to 0.93; $p < 0.001$). Comparisons between Cobb angle and all the other measured parameters failed to show significant correlations (fig. 8, p. 78).

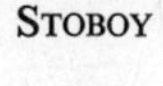

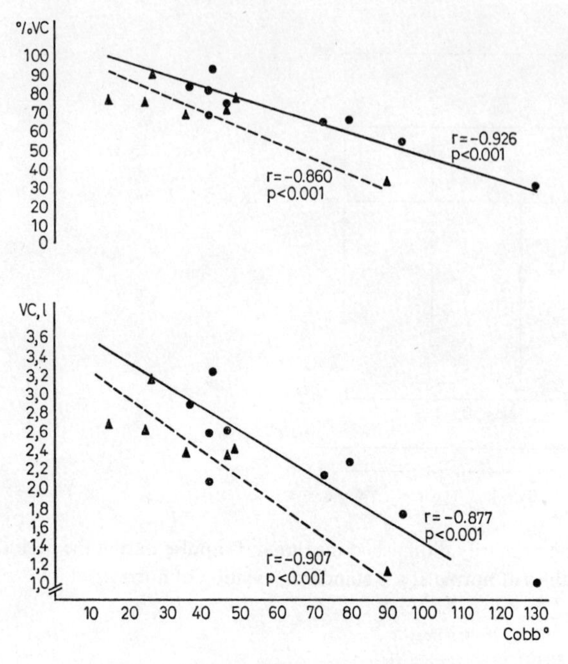

Fig. 8. Relations between predicted vital capacity (%) and absolut values of vital capacity (1) respectively and the Cobb angle at admission and the end of the observation period. ●—●—● = admission; ▲---▲---▲ = end of observation period.

Discussion

The judgment of these obtained results are not as pessimistic as those in other reports, which rely only on pulmonary volumes and capacities measured during resting conditions. WESTGATE and MOE [1969] considers the preservation of the respiratory function on the pre-treatment level as successful. But he, GUCKER [1962] and SCHEIER [1967], found in several patients an additional decrease of VC (10–26%). Only MEZNIK *et al.* [1972] observed a significant increase of vital and total capacity ($p < 0.05$). In these investigations, VC and FVC remained at the preoperative level, but the maximum voluntary ventilation was markedly enhanced [MEZNIK *et al.*, 1972; SCHEIER, 1967]. A pronounced decrease of FVC could not be detected. Also, NAGY and BARTA [1968] found only in some patients a FVC smaller than 70%.

On the other hand, these results are not as optimistic as those of SÜNRAM *et al.* [1974] who, after 4 weeks of ergometric exercises without special treatment, reported a distinct response to training with a significant rise in oxygen consumption, an increase in maximum O_2 pulse and a drop in the respiratory equivalent (patients with Cobb values between 61 and 132°).

In spite of the increased minute volume of respiration, an alveolar hypoventilation can exist due to the increased functional dead-space [Rossier *et al.*, 1958; Ulmer *et al.*, 1970], which can be raised especially at high respiratory frequencies [Ulmer *et al.*, 1970].

Because of the increase in maximum O_2 uptake and O_2 pulse, an improvement of the alveolar ventilation should have been expected. Sünram *et al.* [1974] related the enhanced O_2 pulse after training (8.3–10.4) to an increased alveolar ventilation due to the diminished minute volume of respiration and respiratory equivalent at same workloads. But in this study, no significant changes in maximum minute volume of respiration, respiratory frequency and respiratory equivalent were found.

Some authors [Westgate and Moe, 1969; Aeppli, 1969] reported a rise in O_2 saturation of approximately 2–4%. Heine and Meister [1972] and Sünram *et al.* [1974] measured an increase of the primarily diminished arterial PO_2 (79.5 Torr) especially during or immediately after ergometric exercise (approximately 9 Torr), which was normalized according to age [Lange and Hertle, 1968]. Only Makley *et al.* [1968] and Meznik and Kummer [1970] found in some patients a rise of the arterial PCO_2. In restricted ventilation, the necessary breathing work can usually be efforded [Ulmer, 1970] to keep the arterial PCO_2 in a normal range (37–43 Torr), as reported by Meister and Heine [1973] and Sünram *et al.* [1974]. Shannon *et al.* [1970] as well as Hilpert and Bilge [1973] detected in their investigations with Xenon-133 in scoliotic patients large underperfused pulmonary districts. In these cases with proved restriction and partial hypoventilation, the hypoxemia cannot be completely compensated.

Shannon *et al.* [1970] reported that approximately 40% of the pulmonary capillary blood flow was not arterialized in scoliotic patients under resting conditions. These findings explain the low resting values of arterial PO_2. The decreased pulmonary function is at least partially due to a completely sedentary life, so that presurgical exercise in scoliotic patients is recommendable [Sünram *et al.*, 1974].

The combined surgical and exercise treatment may be able to improve this uneven distribution of perfusion and ventilation and lead to an increase of maximum O_2 uptake and maximum O_2 pulse. Physical exercise should be continued after rehabilitation, to avoid general insufficiency and pulmonary hypertension.

References

AEPPLI, U.: Das Ergebnis der Spondylodese bei Skoliosen Jugendlicher im Hinblick auf die Lungenfunktion. Arch. Orthop. Unfall Chir. *56:* 155–165 (1964).

BERGOFSKY, E.H.; TURINO, F.M., and FISHMAN, A.P.: Cardiorespiratory failure in kyphoscoliosis. Medicine *38:* 236–317 (1959).

BOAS, E.P.: The cardiovascular complications of kyphoscoliosis with a report of a case paroxysmal auricular fibrillation in a patient with severe scoliosis. Am. J. med. Sci. *166:* 89–95 (1923).

CHAPMAN, E.M.; DILL, D.B., and CRAYBIL, A.: The decrease in functional capacity of the lungs and heart resulting from deformities of the chest: pulmonocardiac failure. Medicine *18:* 167 (1939).

COLLINS, D.K. and PONSETT, J.: Long-term follow-up of patients with idiopathic scoliosis not treated surgically. J. Bone Jt Surg. *51A:* 425–455 (1969).

EULER, U.S.: Physiologie des Lungenkreislaufs. Verh. dt. Ges. KreislForsch. *17:* 8 (1951).

GUCKER, P.: Changes in vital capacity in scoliosis. J. Bone Jt Surg. *44A:* 469–481 (1962).

HEINE, J. und MEISTER, R.: Quantitative Untersuchungen der Lungenfunktion und der arteriellen Blutgase bei jugendlichen Skoliotikern mit Hilfe eines 'Funktionsdiagnostischen Minimalprogramms'. Z. Orthop. *110:* 56 (1972).

HILPERT, P. und BILGE, M.: Die Lungenfunktion bei Kyphoskoliose. Technik Medizin *6:* 191–193 (1973).

LANGE, H.J. und HERTLE, F.H.: Zum Problem der Normalwerte; in HERTZ Begutachtung von Lungenfunktionsstörungen (Thieme, Stuttgart 1968).

MAKLEY, J.T.; HERNDON, C.D.; INKLEY, S.; DOERSHUK, C.; MATTHEWS, L.W.; POST, R.J., and LITTLE, A.S.: Pulmonary function in paralytic and nonparalytic scoliosis before and after treatment. A study of sixty-three cases. J. Bone Jt Surg. *50A:* 1379–1390 (1968).

MANKIN, H.J.; GRAHAM, J.J., and SCHACK, J.: Cardiopulmonary function in mild and moderate idiopathic scoliosis. J. Bone Jt Surg. *46A:* 53–62 (1964).

MEISTER, R. und HEINE, J.: Vergleichende Untersuchungen der Lungenfunktion bei jugendlichen Skoliosepatienten vor und nach Operation nach Harrington. Z. Orthop. *111:* 749–755 (1973).

MEZNIK, F. und KUMMER, F.: Skoliose und Lungenfunktion. Z. Orthop. *108:* 382–394 (1970).

MEZNIK, F.; KOLLER, H. und KUMMER, F.: Die Entwicklung der Lungenfunktion nach Skolioseoperationen. Z. Orthop. *110:* 542–544 (1972).

NAGY, G. und BARTA, G.: An skoliotischen Kranken unternommene Atmungsfunktionsuntersuchungen. Z. Orthop. *105:* 166–171 (1968).

ROSSIER, P.H.; BÜHLMANN, A. und WIESINGER, K.: Physiologie und Pathophysiologie der Atmung; 2. Aufl. (Springer, Berlin 1958).

SCHEIER, H.: Prognose und Behandlung der Skoliose (Thieme, Stuttgart 1967).

SHANNON, D.C.; RISEBOROUGH, E.J.; VALENCA, L.M., and KAZEMI, H.: The distribution of abnormal lung function in kyphoscoliosis. J. Bone Jt Surg. *52A:* 131–144 (1970).

SÜNRAM, F.; GÖTZE, H.G. und SCHEELE, K.: Arterielle Blutgase und Säure-Basen-Verhältnisse nach dosierter Ergometerbelastung bei 12–18jährigen Mädchen mit idiopathischen Thorakalskoliosen vor und nach einem vierwöchigen Training. Sportarzt Sportmed. *25:* 6–12, 33–38 (1974).

ULMER, W.T.; REICHEL, G. und NOLTE, D.: Die Lungenfunktion. Physiologie u. Pathophysiologie Methodik (Thieme, Stuttgart 1970).

ULMER, W.T.; REIF, E. und WELLER, W.: Die obstruktiven Atemwegserkrankungen (Thieme, Stuttgart 1966).

WESTGATE, H.K. and MOE, J.H.: Pulmonary function in kyphoscoliosis before and after correction by the Harrington instrumentation method. J. Bone Jt Surg. *61A:* 935–945 (1969).

Dr. H. STOBOY, Orthopädische Klinik und Poliklinik der Freien Universität Berlin im Oskar-Helene-Heim und Institut für Leistungsmedizin Berlin, Clayallee 229, *1000 Berlin 33*

Medicine Sport, vol. 11, pp. 82–88 (Karger, Basel 1978)

Supine Bicycle Exercise in Pediatric Cardiology

GORDON R. CUMMING

Department of Cardiology, Health Sciences Children's Centre,
Winnipeg, Mannitoba

Very few heart patients have symptoms at rest, and this is particularly true in children with congenital lesions who have survived infancy. Children with significant defects are relatively asymptomatic during childhood, and yet in many surgery is advised to prevent potential difficulties in later years. To assess the heart's function as a pump during vigorous exercise in children with congenital defects, exercise studies involving heart catheterization were done on 150 children, 5–16 years of age. This preliminary report details the findings in a group of children with no significant defect, and in those with four types of obstructive lesions.

Methods

Catheters were inserted in the arm veins percutaneously or by cut-down and advanced to the pulmonary artery. The brachial artery at the elbow was cannulated with a Cournand needle. In some children, the femoral vein was catheterized as there was no suitable superficial arm vein. In patients with aortic stenosis, the left ventricle was catheterized by retrograde aortic catheterization after percutaneous insertion of a No. 6F teflon catheter into the femoral artery. The subjects were able to perform leg exercise while the catheters were in the femoral artery or vein, although a few had to stop because of claudication in the catheterized side. Distal pulses were satisfactory in all cases after the catheter was removed. This study is restricted to supine bicycle exercise. A few subjects have been exercised sitting upright on the ergometer, but the problems of looking after the catheters in the mobile upright child and maintaining sterility, although not unsurmountable, do require extra effort. The Elema No. 369 ergometer which provides a reasonably constant load at cycling rates of 50–70 rpm was used throughout, and the ergometer was calibrated at least once yearly [CUMMING and ALEXANDER, 1968]. The cardiac output was measured by the indicator dilution method using indocyamine green dye, a Waters dichromatic densitometer, and a Hewlett-Packard No. 3000 cardiac output computer. This system was calibrated using the dynamic method [SHINEBOURNE *et al.*, 1967] during one of the rest periods between exercises.

Repeat calibrations were obtained before, and at the end of intense exercise in three subjects, and showed no difference, so that only single calibrations midway through the study were used in the other tests.

The resting measurements were obtained about 15 min after insertion of the catheters, and about 5 min after placing the subjects' feet on the ergometer. The subjects pedalled for about 3 min at a work load of 5–8 kpm/kg, rested 3 min, and then pedalled again for 3 min at a work load of about 10–12 kpm/kg. After a 5- to 10-min rest period, they then pedalled at a work load of 15–25 kpm/kg for as long as they could before exhaustion, usually 60–90 sec. Leg fatigue was the reason for stopping exercise in all subjects. For the first two loads, the subjects cycled at 60 rpm, for the final load 70–75 rpm. Indicator curves for measuring cardiac output were obtained during the last 30 sec of exercise, and also at 20, 70 and 120 sec after the exercise had stopped.

Subjects

Over the past 10 years, it has been possible to obtain data on 39 boys and 29 girls who could be classified as having normal hearts. After diagnostic studies were completed, it was concluded that 14 of these had innocent murmurs, 10 had very mild pulmonary stenosis with RV to PA peak systolic gradients of less than 15 mm at rest and 30 mm with exercise, and the remainder had catheters inserted for other reasons such as abdominal, cerebral or pulmonary angiography, and for conditions that were not debilitating and likely to interfere with exercise capacity. The subjects had no prior experience with supine bicycle exercise. Most had been previously tested on a similar ergometer using upright exercise monitoring heart rate only. The subjects were fasting and were sedated with 1 mg/kg of pethedine to a maximum of 25 mg.

Results

For the normal group, the mean values for cardiac index and stroke volume index are summarized in table I. There was no significant sex difference for resting cardiac index. The mean values are not high considering the anxieties created by the test procedure and the posture, supine with the feet slightly elevated on the pedals. Stroke volume index was 15% higher in boys than in the girls.

Steady state exercise led to a variable increase in stroke volume. Using the highest stroke volume recorded during 'steady state' exercise (one to three loads per subject), the main change in stroke index was +9 ml/beat/m^2 (13%) for boys, and +7 ml/beat/m^2 (14%) for girls.

The highest stroke indices were invariably recorded during recovery after exercise. The mean peak value for boys was 72 and girls 65 ml/beat/m^2 of body surface area. This was an increase of about 39% above resting, and 23%

Table I. Cardiac index and stroke volume in 'normal' subjects during rest and exercise

	Boys	Girls	*t*
Resting cardiac index, l/min/m²	3.92 ± 0.20	4.00 ± 0.74	0.4
Resting SVI, ml/beat/m²	51.9 ± 5.9	45.3 ± 6.3	4.1
Highest steady state exercise SVI, ml/beat/m²	58.8 ± 9.7	51.8 ± 8.2	3.0
Highest recovery SVI, ml/beat/m²	72.2 ± 11.0	64.6 ± 11.3	2.6
Peak cardiac index, l/min/m²	10.1 ± 1.75	8.6 ± 1.81	3.4
Maximal heart rate, beats/min	170 ± 17	174 ± 11	1.0
SVI during maximal work, ml/beat/m²	55.7 ± 12.7	46.3 ± 3.1	3.1
Age, years	12.6 ± 3.5	11.8 ± 3.1	–
Number of subjects	31	29	–

SVI = Stroke volume index.

Table II. Increase in cardiac output above resting for various 'steady-state' work loads, normal subjects

Work load kpm/min	Ages 5–16 years		Ages 5–9 years	
	n	$\Delta \dot{Q}$ l/min	n	$\Delta \dot{Q}$ l/min
100	10	2.06 ± 0.85	6	1.93 ± 0.73
170	17	2.93 ± 0.65	12	2.90 ± 0.51
220	3	2.46 ± 1.06	3	2.46 ± 1.06
276	32	3.63 ± 1.16	7	4.30 ± 0.78
333	11	3.28 ± 0.96	6	3.41 ± 1.03
390	16	5.11 ± 1.07	5	5.27 ± 0.45
510	16	5.94 ± 1.74	5	5.72 ± 10.3
550	3	6.47 ± 2.12	0	–
630	14	5.94 ± 1.74	0	–
750	9	8.83 ± 2.36	0	–

All subjects: $\Delta \dot{Q} = 1.03 + 0.00946$ work load in kpm/min, SD = 1.34, r = 0.80. 5- to 9-year-olds: $\Delta \dot{Q} = 1.19 + 0.00901$ work load in kpm/min, SD = 0,87, r = 0.79.
$\dot{Q}$ = Cardiac output in l/min. Δ = Increase in cardiac output above resting, i.e. $\dot{Q}$ exercise – $\dot{Q}$ rest.

above the peak steady state exercise values. This interesting finding was the subject of a previous report from this laboratory [CUMMING, 1972].

The maximal heart rates were about 25 beats/min below that found in the same subjects exercised upright on a previous occasion with the bicycle ergometer. During maximal work when the subjects were straining, stroke

Table III. Mean values, cardiac index, stroke volume index, normals and selected abnormals

	N boys	N girls	PS boys	PS girls	AS boys	HOCM	COA
Resting CI, $l/min/m^2$	3.9 ± 0.2	4.0 ± 0.7	3.5 ± 0.8	3.8 ± 0.8	4.0 ± 0.9	4.1 ± 1.3	4.3 ± 0.5
Resting SVI, $ml/beat/m^2$	52 ± 6	45 ± 6	41 ± 5	39 ± 6	45 ± 9	50 ± 10	45 ± 8
Highest ex CI, $l/min/m^2$	10.0 ± 1.7	8.6 ± 1.8	8.2 ± 1.2	7.2 ± 1.5	8.9 ± 2.1	8.3 ± 1.6	7.8 ± 1.0
Max HR	170 ± 17	174 ± 11	169 ± 12	175 ± 12	171 ± 12	152 ± 20	170 ± 11
Highest recovery SVI	72 ± 11	65 ± 11	62 ± 12	51 ± 5	64 ± 11	63 ± 8	59 ± 11
Number of subjects	31	29	8	8	13	5	9
Mean age, years	12 ± 3	12 ± 3	9 ± 4	8 ± 3	9 ± 2	14 ± 3	10 ± 3

CI = cardiac index; SVI = stroke volume index; ex = exercise; HR = heart rate; N = normal; PS = pulmonary stenosis; AS = aortic stenosis; HOCM = hypertrophic obstructive cardiomyopathy; COA = coarctation of aorta.

volume tended to be a little less than the value observed during previous steady-state exercise.

In relating the cardiac output to the work load for the subjects of different size with different resting cardiac outputs, the best results were obtained by subtracting the resting cardiac output from the exercise output. These results are summarized in table II. No definite sex difference was observed. The values for the younger subjects (age 5–9 years) were comparable to the entire group and regression equations for each group were similar. Cycling against zero frictional load would be expected to increase the cardiac output above resting and the positive Y intercept confirms this in the regression equations. The minimum load setting on the ergometer system in use, taking friction into account, was 100 kpm/min.

Table III lists mean values for normal subjects, and those with pulmonary stenosis, aortic stenosis, hypertrophic obstructive cardiomyopathy and coarctation of the aorta. Under resting conditions, cardiac index tended to be slightly lower in the male patients with pulmonary stenosis but not in the other groups. The mean resting stroke volume index was also 10–20% lower in the patients, with the exception of those with obstructive cardiomyopathy.

The mean of the peak values for cardiac index during exercise was low in all the patient groups although there was considerable overlap. There was no difference between the patients and the normals for maximum heart rate. The highest recovery stroke volume index also tended to be lower in the abnormal groups. All of these patients were asymptomatic and their parents denied any effort intolerance.

Discussion

In the diagnostic heart catheterization laboratory, oxygen uptake ($\dot{V}_{O_2}$) is not an easily measurable parameter in young children either at rest or during exercise, for considerable attention needs to be paid to small details in collecting expired air. More than lip service should be given to the steady-state requirement of the Fick principle when oxygen uptake and A-V oxygen differences are used to calculate cardiac output [VISCHER and JOHNSON, 1953]. While not without its own drawbacks, the indicator dilution method is easier to use because complete air collection is not required, the measurement requires only 10 sec so that steady state is not mandatory, and the duration of exercise can therefore be shorter. With indicator method, maximal exercise measurements where a steady state is not obtained are possible.

The fit adult has difficulty in telling the difference between 200 and 400 kpm/min work loads, while to the child of 15 kg, this is the difference between a load he can sustain for 10 min and a load he can barely cycle against for a minute. A satisfactory correlation between cardiac output and work load, even at these small work loads, indicates that work load is a reasonable substitute for $\dot{V}_{O_2}$ as the reference source for exercise studies.

The efficiency of steady-state bicycle ergometer exercise work of 300–1,500 kpm/min in children is about 23% [CUMMING, unpublished]. Unpublished data from this laboratory, using 11–14-year-old boys as subjects, indicate that there was no difference between the $\dot{V}_{O_2}$ requirement for supine and upright bicycle exercise with the same submaximal work loads of about 4, 8 and 12 kpm/min/kg body weight.

The maximal heart rates were well below those reported for normal children. Maximum heart rates during supine exercise are known to be 10–15 beats/min lower in healthy young adults [STENBERG *et al.*, 1967]. Values in young children are not available. In an unpublished study in this laboratory, previously referred to, normal children showed maximal heart rates 9 ± 6 beats/min lower for supine compared to upright bicycle exercise. Another factor leading to the slow maximal heart rates in the present study is the short duration of exercise. Motivation was sometimes difficult to achieve in the unselected patient coming to heart catheterization, so that short duration work periods were tolerated better than longer periods. Motivation was required for the subject's maximal work regardless of exercise duration, and this is a difficult factor to assess. The mechanical problem facing the young child to get enough leverage to cycle against a stiff braking force while lying supine and having one arm fixed out on an arm board and not available for hanging on, was an added factor. Technicians supported the shoulder where necessary.

The value for maximal cardiac output in normal boys was 10 $l/min/m^2$, quite similar to those reported for well-trained young men by STENBERG (11.3 $l/min/m^2$). With the differences in training, motivation and methodology, larger differences might have been expected.

The cardiac output/work load relationship in our subjects compared favorably to available values in the literature for young adults. The regression line in these subjects lay between values calcultated from the reports of STENBERG *et al.* [1967] and BEVEGARD *et al.* [1966].

The main object of this report was to provide a normal range for the hemodynamic response to exercise in relation to work load in normal children. Exercise studies, including maximal exercise studies, can be obtained in the diagnostic heart catheterization laboratory in the majority of children over

5 years of age, and this includes those with congenital heart defects. In patients with obstructive lesions, there was no consistent relationship between the severity of the obstruction and the impairment in cardiac output or stroke volume. In adult subjects with pulmonary stenosis, there is some impairment in exercise cardiac output with a failure to return to normal after cardiac surgery [JONSSON and LEE, 1968; JOHNSON, 1962; IKKOS *et al.*, 1966; JONSSON, 1969]. Whether operation during early childhood can lead to normal function remains to be determined.

None of the subjects included in this report were restricted in any way from participation in sports or physical education, although self-imposed restrictions might have occurred, or some parental anxieties might have been transferred to the subjects, reducing their participation in physical activities. It can only be presumed that the impaired maximal cardiac output and stroke volumes were the result of cardiac lesion or lesions rather than the result of reduced everyday physical activity.

References

BEVEGARD, S.; FRYSCHUSS, U., and STRANDELL, T.: Circulatory adaption to arm and leg exercise in supine and sitting position. J. appl. Physiol. *21:* 37–46 (1966).

CUMMING, G.R.: Unpublished data.

CUMMING, G.R.: Stroke volume during recovery from supine bicycle exercise. J. appl. Physiol. *32:* 575–578 (1972).

CUMMING, G.R. and ALEXANDER, W.D.: The calibration of bicycle ergometers. Can. J. Physiol. Pharmacol. *46:* 917–919 (1968).

IKKOS, D.; JONSSON, B., and LINDERHOLM, H.: Effect of exercise in pulmonary stenosis with intact ventricular septum. Br. Heart J. *28:* 316 (1966).

JOHNSON, A.M.: Impaired exercise response and other residua of pulmonary stenosis after valvotomy. Br. Heart J. *24:* 375 (1962).

JOHNSON, B. and LEE, S.J.K.: Hemodynamic effects of exercise in isolated pulmonary stenosis before and after surgery. Br. Heart J. *30:* 60 (1968).

JONSSON, B.: Effects of exercise in aortic stenosis before and after surgery; in DENOLIN Rehabilitation of non-coronary heart disease, p. 93 (1969).

SHINEBOURNE, E.; FLEMING, J., and HAMER, J.: Calibration of indicator dilution curves in man by the dynamic method. Br. Heart J. *29:* 920 (1967).

STENBERG, J.; ASTRAND, P.O.; EKBLOM, B.; ROYCE, J., and SALTIN, B.: Hemodynamic response to work with different muscle groups sitting and supine. J. appl. Physiol. *22:* 61–70 (1967).

VISSCHER, M.B. and JOHNSON, J.A.: Fick principle analysis of potential errors in its convention application. J. appl. Physiol. *5:* 635–638 (1953).

G.R. CUMMING, MD, Department of Cardiology, Health Sciences Children's Centre, 685 Bannatyne Avenue, *Winnipeg, Manitoba R3E OW1* (Canada)

Medicine Sport, vol. 11, pp. 89–95 (Karger, Basel 1978)

Working Capacity in Anorexia Nervosa[1]

CLAES THORÉN

St. Göran's Children's Hospital, Stockholm

Primary anorexia nervosa (AN) occurs mainly in young females and is characterized by self-enforced food avoidance, resulting in starvation and weight loss with extreme emaciation. It usually occurs in pubescent girls and more rarely in late childhood and in males. AN is often seen as a phobic avoidance response to the physical changes of maturation and involves an extensive disturbance in personality development and separation-individuation. The psychodynamic and clinical features of AN and the extensive psychiatric and psychosomatic literature are reviewed by BRUCH [1965] and ROWLAND [1970]. In monographs by DALLY [1969], THEANDER [1970] and THOMÄ [1961], the history and prognosis of this psychosomatic illness is presented.

From a physiological point of view, the basic problem is their lack of energy input and numerous adaptive changes characterize the AN syndrome such as hypothermia, reduced basal metabolic rate, dry skin, bradycardia, bradypnea, hypotension and acrocyanosis. These findings seem to be secondary to the adaption to chronic inanition. In mature girls, amenorrhea is an early and cardinal effect of starvation in AN. Growth is retarded, which if it occurs early may become a major problem. The abnormalities in endocrine function are believed to be secondary to hypothalamic dysfunction [LUNDBERG *et al.*, 1972].

ECG changes in AN frequently consist of low voltage, flattened or inverted T waves and depressed S-T segments as well as arrhythmias [SILVERMAN, 1974]

[1] This study was supported by the Foundations of Harald and Greta Jeansson and Axel Tielman.

and are influenced by exercise and pharmacological drugs [FOHLIN and THORÉN].

The patients are often overactive and show a conscious effort to increase their energy output in line with their active resistance to calories. Their output may be very large, and difficult to assess ritualistic activities, walking with dogs and other energy-consuming physical activities without complaints of fatigue. A complexity of compensatory factors may help them to be physically hyperactive.

An increased interest in AN is evidenced by multidisciplinary publications in recent years [ROWLAND, 1970; REICHSMAN, 1972]. Cardiac and circulatory function has, however, been largely neglected despite the bradycardia and ECG changes and the fact that physical activity forms a part of the AN syndrome. These patients may help to explain how physiological adaption solves the demand for muscular exercise in isolated caloric starvation.

By examining the working performance and the reaction of the heart and circulation to heavy and maximal exercise in patients with AN, this presentation is a preliminary report of studies in progress.

Material and Methods

The series consisted of 20 patients diagnosed by the Child Psychiatric Clinic. There were 13 girls with a mean age of 16 years (r: 12.5–20) and 7 boys with a mean age of 13.6 years (r: 10–16). The body weight averaged 37.6 kg (r: 24.6–50) for girls and 39.2 (r: 28–56) for boys with a mean deviation in relation to the height of −2.3 and −2.1 SD, respectively, corresponding to a weight loss of about 25–30% of normal average.

Exercise tests were performed on a bicycle ergometer with stepwise increase of work loads up to exhaustive level. The ECG and blood pressure were recorded during the test. Expired air was collected and samples were analyzed by the micro-Scholander method for determination of oxygen uptake. Blood lactate was measured enzymatically at rest, during and after exercise.

Results

The results showed no difference between girls and boys concerning heart and respiratory rate, blood pressure and blood lactate. The mean heart rate at rest was 53 beats/min (r: 37–88). During exercise, the heart rate increased linearly to work load (fig. 1) but was mostly high in relation to O_2 uptake (fig. 2). The maximal heart rate was low with a mean of 170 beats/min (r: 142–193).

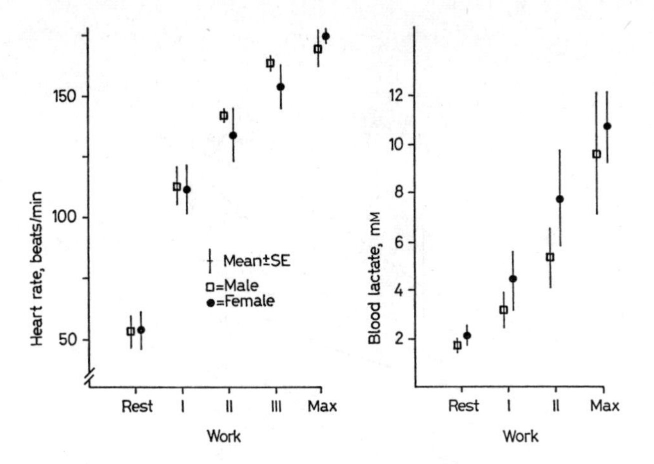

Fig. 1. Mean heart rate and blood lactate at rest and during stepwise increasing work loads.

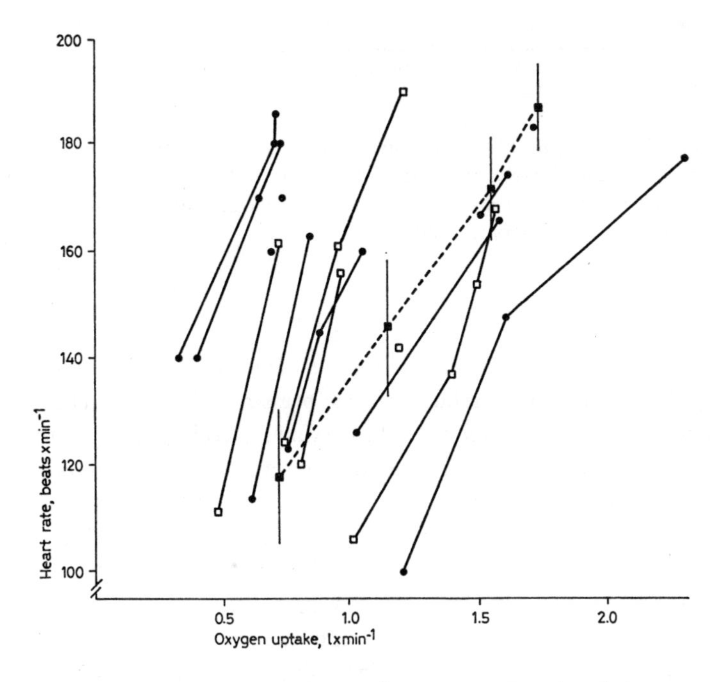

Fig. 2. Individual heart rate of anorexia nervosa patients in relation to oxygen uptake during exercise with increasing work load up to exhaustion. Filled symbols denote girls. Dashed line denotes the mean values with ±1 SD of 11- to 13-year-old healthy boys [ERIKSSON, 1972].

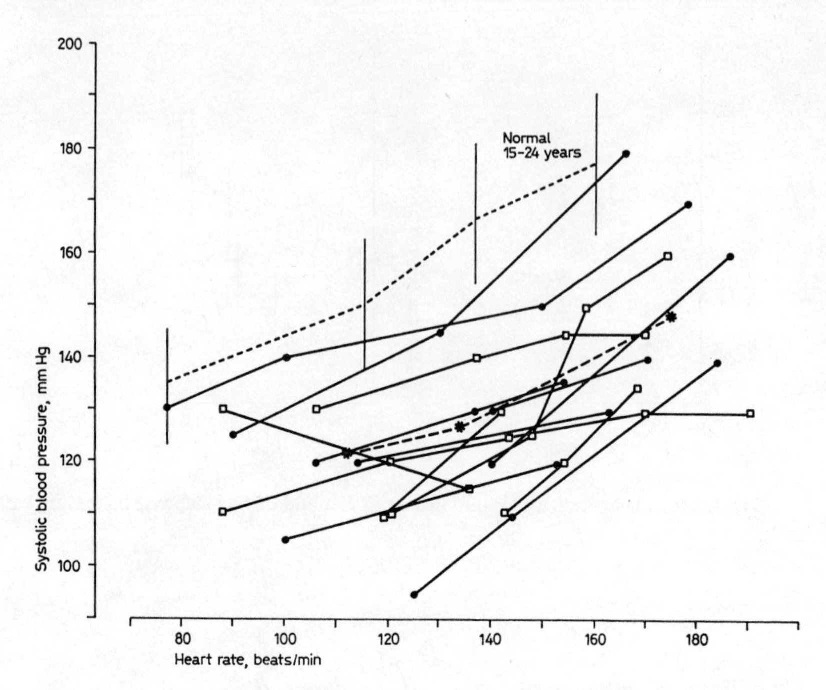

Fig. 3. Individual systolic blood pressure during stepwise increases in exercise to maximal level in anorexia nervosa in relation to heart rate. Dashed line between stars indicates mean values. The upper dashed line denotes the mean values ±1 SD of 15- to 24-year-old healthy men [KARLEFORS, 1966].

The systolic blood pressure was low at rest (mean: 102/70) and during exercise (fig. 3) with a mean maximal systolic pressure of 147 mm Hg. Only 2 girls reached values over 160 mm Hg at maximal exercise. The respiratory rate was low at rest 14.9 (r: 8–21) and reached a mean maximal value of 36.6/min (r: 24–62).

Blood lactate was generally somewhat high, even at rest, with a mean value of 1.88 mM/l (r: 0.64–3.14) and increased more or less lineraly with the work load (fig. 1) up to a maximal mean value of 9.84 mM. There was a tendency for higher values in girls but without a significant difference. One 16-year-old boy reached an extremely high value of 23.0 mM/l. Analysis of muscle biopsies at the same time also showed a correspondingly high muscle lactate concentration which may indicate a high anaerobic energy supply. Maximal oxygen uptake in relation to body weight showed a rather

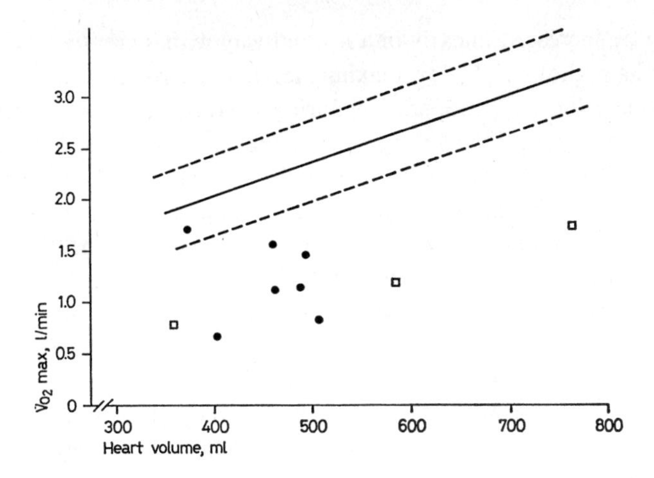

Fig. 4. Maximal oxygen uptake in relation to heart volume in 10 AN patients. The regression line ±250 of healthy children.

wide range from very low (23 ml) to normal (49 ml). The mean value was 33 ml/kg. In relation to heart volume the max. Og uptake was far below expected normal range (fig. 4).

Discussion

All of the cases showed certain clinical changes typical for AN and secondary to the nutritional deficiency. The bradycardia was more pronounced in this series than in those reported by SILVERMAN [1974] of similar age. On the other hand, his patients showed a more marked bradypnea and hypotension. Despite all the physiological and metabolic abnormalities associated with AN, there are no studies of ergometry reported.

The heart rate increased linearly in a normal way with increasing work load but reached a pathologically low maximal level. These patients suffering from AN were usually very cooperative during exercise tests and surely reached an exhaustive level. The leveling off criterion was also fulfilled in 8 of the 16 analyzed patients. The high maximal blood lactate speaks also in favor of a maximal exercise with a large anaerobic energy supply. The AN patients showed values similar to well-trained normal subjects of the same age [ERIKSSON, 1972; THORÉN *et al.*, 1973]. Elevated lactate levels at rest and the rapid increase during work in contrast to the normally non-linear rise might be explained by a metabolic adaption resulting in an increased anaerobic energy capacity than normal. ENRICH and BAETHKE [1971] described a 29-year-old woman with a chronic AN syndrome with vomiting and hypopotasemia who had an elevated blood lactate (25 mg%) at rest and showed an increase to 65 mg% at a low work load (1 W/kg).

The abnormally low systolic blood pressure found even during heavy exercise may be a part of the compensatory mechanism for conserving energy and reducing the load on the cardiac muscle.

The maximal oxygen uptake in AN, which must be the most correct way of determining the working capacity, was generally low even in relation to the low body weight.

Typical primary anorexia nervosa certainly occurs in males [BEUMONT *et al.*, 1972] and this study also shows that the cardiac findings and exercise function in boys closely resemble those in the female.

Further studies of the hemodynamics and functional dimensions, the peripheral circulation and temperature regulation, the metabolism during prolonged exercise and muscular fibers and metabolism in AN are planned to elucidate the many questions to which this preliminary study has given rise.

References

BEUMONT, P.J.V.; BEARDWOOD, C.J., and RUSSELL, G.F.M.: The occurrence of the syndrome of anorexia nervosa in male subjects. Physiol. Med. *2:* 216–231 (1972).

BRUCH, H.: Anorexia nervosa and its differential diagnosis. J. nerv. ment. Dis. *141:* 555–566 (1965).

DALLY, P.: Anorexia nervosa (Grune & Stratton, New York 1969).

ENRICH, H.M. und BAETHKE, R.: Überproduktion von Lactate bei Anorexia nervosa. Klin. Wschr. *49:* 501 (1971).

ERIKSSON, B.O.: Physical training, oxygen supply and muscle metabolism in 11–13-year-old boys. Acta physiol. scand. suppl. 384 (1972).

FOHLIN, L. and THORÉN, C.: ECG changes in anorexia nervosa (to be published).

KARLEFORS, T.: Circulatory studies in male diabetics. Acta med. scand. *180:* suppl. 449 (1966).

LUNDBERG, P. O.; WÅLINDER, J.; WERNER, J., and WIDE, L.: Effects of thyrotropin-releasing hormone on plasma levels of TSH, FSH, LH and GH in anorexia nervosa. Eur. J. clin. Invest. *2:* 150–153 (1972).

REICHSMAN, F. (ed.): Hunger and satiety in health and disease. Adv. psychosom. Med. (Karger, Basel 1972).

ROWLAND, C. V., jr. (ed.): Anorexia and obesity. Int. Psychiat. Clin. *7:* 37–137 (1970).

SILVERMAN, J. A.: Anorexia nervosa: clinical observations in a successful treatment plan. J. Pediat. *84:* 68–73 (1974).

THEANDER, S.: Anorexia nervosa. A psychiatric investigation of 94 female patients. Acta psychiat. scand. suppl. 214 (1970).

THOMÄ, H.: Anorexia nervosa (Huber-Klett, Bern 1961).

THORÉN, C.; SELIGER, V.; MÁČEK, M.; VÁVRA, J., and RUTENFRANZ, J.: The influence of training on physical fitness in healthy children and children with chronic diseases. Current aspects of perinatology and physiology in children, pp. 83–112 (Springer, Berlin 1973).

Dr. C. THORÉN, St. Göran's Children's Hospital, *Stockholm* (Sweden)

Medicine Sport, vol. 11, pp. 96–101 (Karger, Basel 1978)

Thermoregulation in Anorexia Patients

C.T.M. DAVIES[1], L. FOHLIN and C. THORÉN
St. Göran's Children's Hospital, Stockholm

Introduction

Anorexia nervosa in children is characterized by weight loss due to a failure to eat. The physical features of the disease have been described by FOHLIN *et al.* [1977]. The major changes in physiological function are secondary to starvation, but although the patient may exhibit a depressed metabolism, bradycardia and hypotension at rest, these are symptoms of which he or she rarely complains. Their major physical concern is one of numbness in the hands and feet and a general feeling of cold. Their extremities are usually vasoconstricted, and acrocyanosed and these features are associated with a low resting core (rectal) temperature and cardiac frequency (fig. 1). The present paper is a preliminary report of the thermal responses of 10 anorexic patients under conditions of exercise in a thermally neutral environment.

Material and Methods

The mean age, height, weight and lean body mass (estimated from skinfold thickness after the method of PARIZKOVA [1961]) were 15 ± 2.0 years; 164 ± 15 cm; 37.9 ± 8.1 kg; and 34.1 ± 7.2 kg, respectively. Nine of the 10 patients were girls and the anorexic state had been present prior to the investigation for a mean period of 1 year (range 0.5–3 years). They were measured at rest and during prolonged exercise (1 h) at 65% of their maximal aerobic power output ($\dot{V}O_{2\,max}$) on an upright stationary bicycle ergometer. Rectal temperature

[1] One leave of absence from Medical Research Council, Environmental Physiology Unit, London School of Hygiene and Tropical Medicine, London.

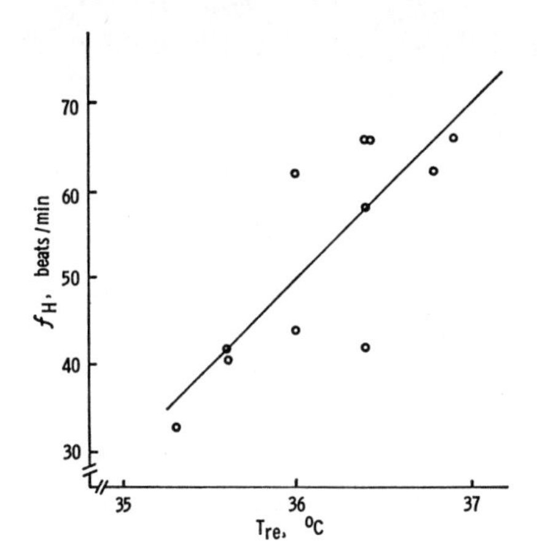

Fig. 1. Resting rectal temperature (T_{re}) and cardiac frequency (f_H) in anorexia nervosa. The data were taken from the ward records of 11 patients admitted to St. Gorän's Hospital for treatment.

(T_{re}) was measured continuously by means of a thermocouple inserted 8 cm above the internal sphincter, and skin temperature (T_{sk}) was determined at 14 sites [HARDY and DUBOIS, 1938] at regular intervals, during exercise. The weighted average of T_{sk} was taken as the mean body skin temperature ($\bar{T}_{sk}$), oxygen intake ($\dot{V}O_2$) was measured by the open circuit technique at the 28th and 58th min of exercise. The loss of weight during exercise was determined by nude weighing on a beam balance (accurate to ± 10 g) immediately before and after exercise. From these raw data, calculations of metabolic (M) and total heat (H) production, evaporative sweat loss (E), heat storage (S) and peripheral heat conductance were made in the standard way [NIELSEN, 1969]. Five healthy children acted as controls and were measured in exactly the same way as the patients. The ambient temperatures in the laboratory were $T_{db} = 23.6 \pm 0.84$ °C and T_{wb} 14.1 $\pm$ 2.2 °C.

Results

The changes of T_{re} in a typical anorexia patient and control subject are shown in figure 2. In the anorexic patients, the T_{re} generally rose more slowly than in the controls to reach a delayed 'plateau' between the 50th and 80th min of exercise. T_{re} was unrelated to M but was associated with relative work load (% $\dot{V}O_{2\,max}$) in 7 of the patients studied. In the remaining 3 patients, the T_{re} points at the 60th min of exercise were outside the expected limits for the two variables [DAVIES *et al.*, 1976]. $\bar{T}_{sk}$ was lower at rest in the patients compared with the controls and increased during exercise; at the 60th min

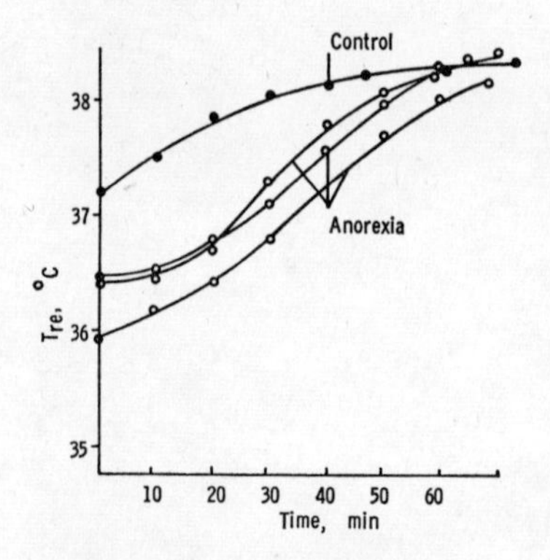

Fig. 2. Changes of rectal (T_{re}) temperature in an anorexic patient (measured on 3 occasions) and a control subject.

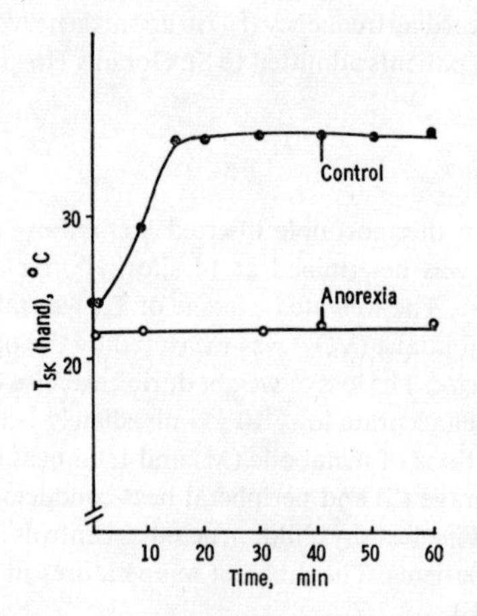

Fig. 3. Hand temperature (T_h) in an anorexic patient and control subject.

there was approximately a 1 °C difference in mean $\overline{T}_{sk}$ in the two groups. However, there was some evidence of selective vasoconstriction for although the $\overline{T}_{sk}$ was higher in the patients, their extremities tended to remain cooler than the controls (fig. 3).

Evaporative sweat loss for a given H lay within the normal limits expected for healthy subjects (fig. 4). The anorexic patients dissipated 25% of their heat by evaporation of sweat from the skin, 10% of their heat was stored and the remaining 65% was lost from the body by convection and radiation. The corresponding figures for the controls were 62%, 7% and 21%, respectively. Heat lost from the lungs corresponded to 9% of H in both groups.

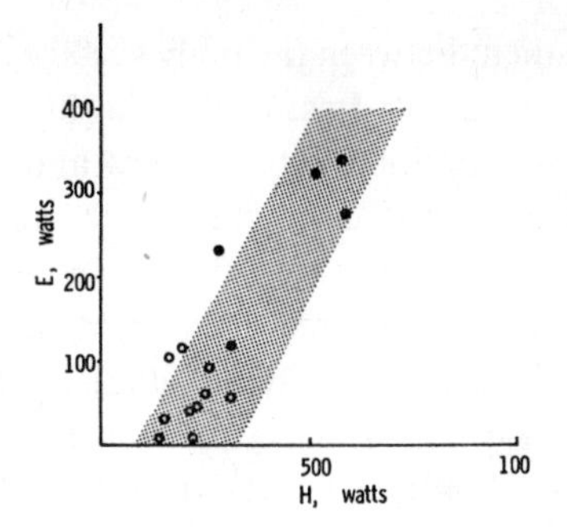

Fig. 4. Evaporative sweat loss (E) in relation to total heat production (H). The shaded area represents the limits of the two variables previously found for healthy young adult subjects [DAVIES *et al.*, 1976]. ○ = Anorexia; ● = controls.

Discussion

Under the laboratory conditions of the present experiments, anorexia nervosa patients appear to regulate their core temperature in a way similar to that found for extremely sedentary and unacclimatised subjects [WYNDHAM *et al.*, 1964]. At rest, the anorexic children have a low T_{re} (fig. 1, 2) and body heat content. This gives rise to a greater capacity for heat storage and may account for the delayed rise in T_{re} observed at the onset of work. The changes in T_{re} at the commencement of work do not appear to affect the final 'plateau' level of T_{re} at the 60th min of exercise, and this would suggest (with the exception of 3 cases) that loss of body fat has only a marginal effect on the 'steady-state' level of T_{re} in anorexia patients working in a thermally neutral environment. We are currently studying [DAVIES *et al.*, to be published] the effects of preheating and those of a cool and hot environment on temperature regulation in anorexia.

However, though our results give no support to the views of MECKLENBURG *et al.* [1974], that anorexia produces a primary hypothalamic disturbance affecting the set-point [HAMMEL, 1968] of the thermoregulatory system, it is nevertheless clear that the disease does effect both qualitative and quantitative changes in the way heat is conserved and dissipated from the body. Undoubtedly, the loss of subcutaneous fat does facilitate the loss of heat from the body and it can be seen from our results that in anorexia a greater proportion (65%) of the total heat production is dissipated by convection (C) and radiation (R) when comparison is made with healthy control (21%) subjects. A decrease in thermal insulation will increase the flow of heat across the skin and the resulting temperature gradient between the body surface and the environment will govern the heat lost by C + R. Loss of heat in this way can only be controlled by adjustments to the peripheral blood flow and this the anorexia patient appears to attempt by selectively increasing vasoconstrictor tone (and thereby reducing $\overline{T}_{sk}$) in the extremities (fig. 4). Thus, the patient limits the size of his 'core' and the 'effective' area which heat can be dissipated from the skin through C + R. Under these circumstances, it has been shown in animals and man that heat can be directly exchanged by a counter-current mechanism [Bazett *et al.*, 1948; Schmidt-Nielson, 1963], since blood is diverted from the superficial to the deep veins which lie in close proximity to the arteries. Heat is conserved by a process of rewarming directly the blood returning to the heart. Further experiments to elucidate the exact relationship of peripheral blood flow, skin temperature and evaporative sweat loss to core temperature regulation would be rewarding in anorexia nervosa patients.

Acknowledgements

We are indebted to Prof. Nylander for allowing us to study his patients, the personnel of the Clinical Physiology Laboratory for their technical assistance and the patients and control subjects for their cooperation in this study. The investigation was supported by grants from the Sigurd and Elsa Gobjés Foundation.

References

Bazett, H.C.; Love, L.; Newton, M., Eisenberg, L.; Day, R., and Forster, R.: Temperature changes in blood flowing in arteries and veins in man. J. appl. Physiol. *1:* 3 (1948).

Davies, C.T.M.; Brotherhood, J., and Zeidi Fard, E.: Temperature regulation during severe exercise with some observations on the effects of skin wetting. J. appl. Physiol. *41:* 772–776 (1976).

Fohlin, L.; Freyschuss, U.; Bjarke, B.; Davies, C.T.M., and Thorén, C.: Function and dimensions of the circulatory system in anorexia nervosa. Acta paediatr. scand. *67:* 11–16 (1978).

Hammel, H.T.: Regulation of internal body temperature. Annu. Rev. Physiol. *30:* 691–710 (1968).

Hardy, J.D. and Dubois, E.F.: The technic of measuring radiation and convection. J. Nutr. *15:* 461–475 (1938).

Mecklenburg, R.S.; Loriaux, D.L.; Thompson, R.H.; Andersen, A.E., and Lipsett, M.B.: Hypothalamic dysfunction in patients with anorexia nervosa. Medicine *53:* 147–159 (1974).

NIELSEN, B.: Thermoregulation in rest and exercise. Acta physiol. scand. suppl. 323 (1969).
PAŘÍZKOVÁ, J.: Total body fat and skinfold thickness in children. Metabolism *10:* 794–807 (1961).
SCHMIDT-NIELSEN, K.: Heat conversation in counter-current systems; in HARDY Temperature: its measurement and control in science and industry, vol. 3, p. 143 (Reinhold, New York 1963).
WYNDHAM, C.H.; STRYDOM, N.B.; MUNRO, A.; MACPHERSON, R.K.; METZ, B.; SCHEFF, G., and SCHIEBER, J.: Heat reactions of Caucasians in temperature, in hot, dry and in hot, humid climates. J. appl. Physiol. *19:* 607–612 (1964).

Dr. C.T.M. DAVIES, MRC Envisonmental Physiology Unit, London School of Hygiene and Tropical Medicine, Keppel Street, *London WC1E 7HT* (England)

Medicine Sport, vol. 11, pp. 102–107 (Karger, Basel 1978)

Body Dimensions and Exercise Performance in Anorexia Nervosa Patients

L. FOHLIN, C.T.M. DAVIES, U. FREYSCHUSS, B. BJARKE and C. THORÉN
St. Göran's Children's Hospital, Stockholm

Anorexia nervosa (AN) is characterized by severe loss of weight due to the patient's active refusal to eat. The aetiology is complex and there is a peculiar combination of psychological and physiological manifestations together with signs of a hypothalamic dysfunction [LUPTON *et al.*, 1976]. Numerous changes in body function do occur secondary to the starvation. Among the most common ones are bradycardia, low blood pressure and body temperature. For hospitalized patients, a mortality rate somewhere around 10% is reported.

Despite of the fact that many of these patients are hyperactive, very little is known about their cardiovascular adaptation to long-standing caloric restriction. The purpose of the present study was to determine different components of the oxygen transporting system in AN and their relations to aerobic capacity and body weight. Cardiac output and stroke volume at rest and during exercise were also studied. This study is a continuation to that presented at Seć, which is published in this volume as an introduction to this paper [THORÉN].

Material and Methods

Between 1971 and 1975, 30 patients with anorexia nervosa, 20 girls and 10 boys, have been examined. All of them fulfilled the diagnostic criteria for AN, and all but 3 patients required hospital care. Some anthropological data of the material are given in table I.

The mean age of the girls was 15.4 years and of the boys 14.3 years. The average weight loss was 26.8% for the girls and 23.5% for the boys and the mean weight and height was 36.9 kg and 162.2 cm for the girls and 38.9 kg and 166 cm for the boys, respectively.

Table I. Mean values with range for some anthropological data in anorexia nervosa patients

	Age years	Weight kg	Height cm	Weight loss %	Duration years
Girls (n = 20)	15.4 (12.1–18.2)	36.9 (24–48.5)	162.2 (132–180)	26.8 (14–40)	1.2 (0.5–3.0)
Boys (n = 10)	14.3 (11.4–18.0)	38.9 (27.3–56)	166 (142–188)	23.5 (21.5–38)	0.9 (0.5–2.0)

Blood volume was determined with ^{125}I-labelled albumin. Heart volume was measured by X-ray with the patients in prone position. Cardiac output determinations were made using the dye-dilution method with indocyanine green. Oxygen uptake was determined at rest and during exercise on a bicycle ergometer (ELEMA) using the Douglas bag technique and gas analyses by micro-Scholander technique. Blood lactate was analyzed enzymatically. For further methodical details and references, see FOHLIN *et al.* [1978].

For determination of the maximal oxygen uptake (max $\dot{V}_{O_2}$) the patients were encouraged to pedal to exhaustion on a voluntary maximal load after step-wise warming up on two submaximal loads. With max blood lactate concentrations >8 mmol/l and RQ values >1.0, the $\dot{V}_{O_2}$ was accepted as maximal. According to this criteria, 7 out of 30 patients failed to complete a maximal exercise test.

Results

Blood volume was correlated to body weight within normal range (fig. 1). Blood volume also showed a significant correlation to heart volume (fig. 2).

Heart volume was significantly related to body weight with a more or less identical regression line with that for healthy school children (fig. 3). Max $\dot{V}_{O_2}$ was also significantly correlated to heart volume but with another slope than for healthy individuals (fig. 4).

Heart rate at rest was 51 (37–75) beats/min and the mean maximal heart rate was 176 (152–190) beats/min without sex difference. Maximal blood lactate was 11.0 mmol/l for girls and 13.8 mmol/l for boys.

Mean oxygen uptake in relation to body weight was 31.9 (22–40) and 33.8 (31–39) ml/kg for the girls and boys, respectively. $\dot{V}_{O_2}$ increased linearly with

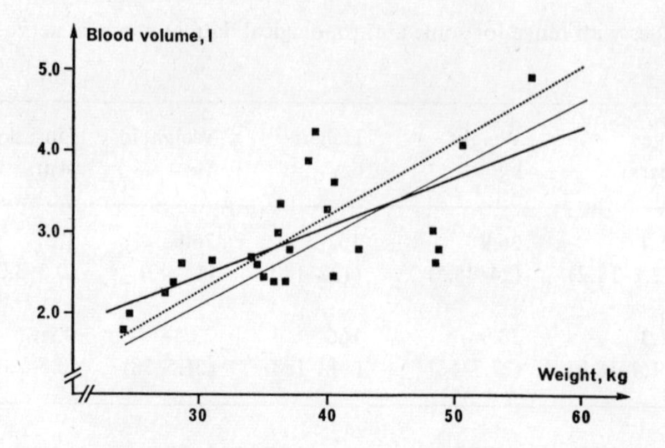

Fig. 1. Individual values for blood volume in relation to body weight in 25 patients with anorexia nervosa. Thick line: regression line (Y = 0.59 + 0.061x; r = 0.69***). Thin and dotted lines: normal regression for girls and boys, respectively [KARLBERG and LIND, 1955].

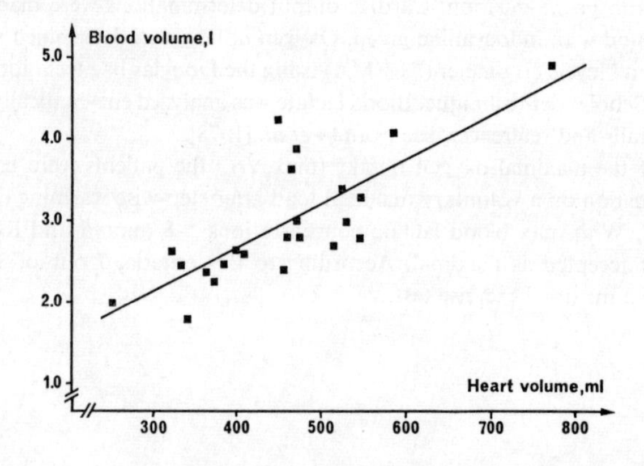

Fig. 2. Individual values for blood volume in relation to heart volume in 23 patients with anorexia nervosa. Regression line is given (Y = 0.48 + 0.01x; r = 0.78***).

work load but on a lower level the downward parallel displacement of the regression line being of the same magnitude as the decrease of the resting oxygen uptake. The latter was also lower than normal in relation to percent of max $\dot{V}_{O_2}$.

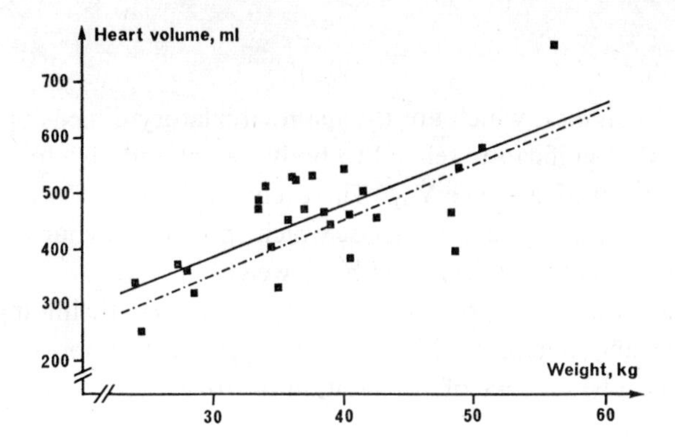

Fig. 3. Individual values for heart volume in relation to body weight in 27 patients with anorexia nervosa. Unbroken line: regression line (Y = 108.11 + 9.29x; r = 0.72***). Broken line: normal regression line (Y = 61.93 + 10.06x; r = 0.86) [THORÉN, 1977].

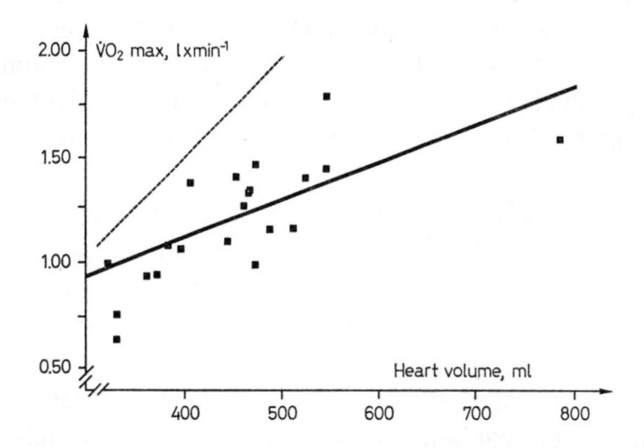

Fig. 4. Maximal oxygen uptake in relation to heart volume in 21 patients with anorexia nervosa. Unbroken line: regression line (Y = 0.398 + 0.002x; r = 0.70***). Broken line: normal regression line (Y = 0.417 + 0.005x; r = 0.774) [THORÉN, 1977].

In 6 patients, the cardiac output was measured at rest and during submaximal work sitting on a bicycle. There was a linear increase with $\dot{V}_{O_2}$ and the values were within normal range. Stroke volume was steady and well maintained during exercise.

Discussion

Heart and blood volumes, which are the main circulatory dimensions, were found to be highly significantly related to body weight and also to the working capacity determined as max $\dot{V}_{O_2}$. These correlations were within the expected range for healthy young subjects, which speaks in favour of a good circulatory adaptation to the decreased body weight in AN.

The low heart rate and blood pressure at rest are characteristic findings in the AN syndrome and together with the low oxygen consumption at rest they all could be phenomena of the adaptation to a hypometabolic state.

Even during exercise, the $\dot{V}_{O_2}$ was low in relation to work load but with the same difference as shown by the oxygen consumption at rest. Maximal $\dot{V}_{O_2}$ in relation to body weight was low, but showed a normal ratio to ventilation volume. The high maximal blood lactate, 11.1 for the girls and 13.8 mmol/l for the boys, speaks very much in favour of a great anaerobic part of the metabolism during the exhaustive exercise. This indicates that the exercise was performed on a maximal level. These high values are more commonly seen in well-trained individuals of the same age. In AN, the leg muscles seem to have an increased anaerobic energy capacity or an increased tolerance to lactate concentration.

Cardiac output was normally related to oxygen uptake and stroke volume was well maintained during work up to submaximal level indicating an unimpaired myocardial function. Thus, the low maximal heart rate is not explained by a cardiac limitation. It could depend on the bradycardia already seen at rest and/or the low body temperature. The exercise performance is, however, only slightly subnormal despite the chronic caloric starvation.

The studies of the thermoregulation in AN are published separately in this volume [DAVIES *et al.*]. The peripheral adaptation of the circulation has also been studied and the results are under publication elsewhere [FREYSCHUSS *et al.*, 1978]. We found a decreased blood flow and a depressed response to in-direct heating as measured by calf blood flow. This circulatory pattern might reflect an appropriate preservation of energy in a hypometabolic state.

References

DAVIES, C.T.M.; FOHLIN, L., and THORÉN, C.: Thermoregulation in Anorexia Patients. Medicine Sport, vol. 11, pp. 96–101 (Karger, Basel 1978).

FOHLIN, L.; DAVIES, C.T.M.; FREYSCHUSS, U.; BJARKE, B., and THORÉN, C.: Function and dimensions of the circulatory system in anorexia nervosa. Acta paediatr. scand. *67:* 11 (1978).

FREYSCHUSS, U.; FOHLIN, L., and THORÉN, C.: Limb circulation in anorexia nervosa. Acta paediatr. scand. *67:* 225 (1978).

KARLBERG, P. and LIND, J.: Studies of the total amount of hemoglobin and the blood volume in children. Acta paediatr., Stockh. *44:* 17 (1955).

LUPTON, M.; SIMON, L.; BARRY, V., and KLAWANS, H.L.: Biological aspects of anorexia nervosa. Life Sci. *18:* 1341 (1976).

THORÉN, C.: Heart volume and aerobic capacity in healthy school children (to be published).

Dr. L. FOHLIN, St. Göran's Children's Hospital, *Stockholm* (Sweden)

Medicine Sport, vol. 11, pp. 108–117 (Karger, Basel 1978)

Methods of Biological Maturity Assessment

MARCEL HEBBELINCK
Vrije Universiteit Brussel, Brussels

From a point of view of an observer's first impression, the most available age is physique age as conveyed through the visual sense. In most instances, physique supplies the information required for an estimate of the status of biological development. Physical and biological changes during the growth process have been described and explained in general terms. In practice, however, observations and experiments are made, for a major part, on individuals of which the growth rates differ. Intra- and interindividual variations are fundamental biological features. These facts of biological variability in growth cause great difficulty, for the ultimate aim of much of biological research is making reliable predictions from orderly relationships among objects or events or characters. The observed phenotype variance of physical development has both genetic and environmental components.

This biological variability and wide variations in the rate at which individuals mature and pass through the various stages of the human life cycle have incited human biologists to search for proper ways of assessing the stage of development. Indeed, wide variations in patterns of physical development, work capacity, and motor proficiency have been observed in children of similar chronological age, and it is obvious that some concept other than chronological age is needed. The concept of changes in biological quality during human growth is generally expressed in terms of biological maturity or shortened biological age, also sometimes called 'developmental age' [STOTT, 1970; TANNER, 1962].

The purpose of this paper is to give a critical survey of the most commonly used methods to assess biological age from birth to maturity.

Basically there are many 'biological, physiological or developmental ages' in each individual, depending on what biological system is being considered. Various data show that different biological functions reach maturity at quite different periods in the life cycle. Commonly four systems to assess biological maturation age have been accepted: sexual maturation age, dental age, morphological size age and skeletal age.

1. Sexual Maturation Age

In 1908, CRAMPTON introduced the term anatomical or physiological age, using the appearence of pubic hair as an indicator of pubescence. Besides secondary sex characters, such as pubic hair, facial and axillary hair, development of genitals, breast development, voice mutation, first ejaculation (oigarche) and menarche have been used in assessing sexual maturation age. During adolescence, these characters express the stage of development quite well. The different stages of development of secondary sex characters have been described by several authors using less or more comparable rating systems [cf. GRIMM, 1966].

Generally spoken, when appraising sexual maturation, investigators in the English-speaking world apply the criteria as indicated by GREULICH *et al.* [1942] or TANNER [1962], whereas in Middle and Eastern Europe, the criteria as defined by ZELLER [1938] or SCHWIDETSKY [1950] are used.

In our cross-sectional epidemiological studies on elementary school children [HEBBELINCK and BORMS, 1975b, 1976] and in our present longitudinal growth project, we have applied the criteria prescribed by TANNER and, for comparative purposes, we used the excellent colour photographs of the different stages of development of pubic hair in boys and girls and of breast development in girls and of genital development in boys, as published by VAN WIERINGEN *et al.* [1971].

Interpretation of the results of the rating and assessment of sexual maturation is not always easy. Pubic and axillary hair are not invariable concomitants of pubescence. On the average, breast development precedes pubic hair growth, but in some 1–5% of girls the converse is true. Moreover, pubic hair is not always curly and highly pigmented, especially in certain ethnic groups [GARN and SHAMIR, 1958]. The size of the penis and the volume of the testes have also been rated or measured, but these appraisals are dependent from the subject's size and emotional factors, particularly in the case of direct measurement using a caliper or more specifically graduated rings [KROGMAN, 1956; REYNOLDS and WINES, 1951] and the orchidometer [PRADER, 1966; VAN WIERINGEN, 1971].

Of all the sexual maturity characters associated with the adolescent years, menarche is the most noteworthy. The onset of the first menstrual flow is a qualitative event of major biological significance, which is easily recordable, if methodological considerations are observed. DE WIJN [1965] has discussed and outlined the most commonly used methods, i.e. (a) *prospective method,* recording onset of periods in a longitudinal study; (b) *retrospective* or *recollection method,* recording age at menarche as stated when interrogated, and (c) *status quo method,* recording whether or not periods have started at the time of the interrogation.

Unequivocally, the prospective method is the most accurate one, but the drawback is that it takes at least 6 years to estimate age of menarche in a population. The retrospective method can be used for comparison of median age at menarche in relatively small groups, while the disadvantages are the poor memory and the tendency of older girls to state higher ages of menarche. Finally, when the prospective method cannot be used, the estimation of the mean age at menarche in population surveys can be calculated by recording whether or not (status quo) menstruation has started. This implies that girls from age 9 or 10 up to 18 must be included in the sample. The drawback of this method is that it requires a large sample of girls.

It is obvious that the criteria of secondary sex characters can be applied only during the period of pubertal development, that is during the age ranges of about 9–16 years. Furthermore, age at menarche is limited to girls.

2. Dental Age

Already as early as 1837, SAUNDERS published an article '*Teeth as a Test of Age*' which he addressed to the members of the British Parliament. He advocated the use of norms for tooth eruption as criteria to limit children's labour.

Eruption or emergence of the teeth from the gums may be observed visually or studied on plaster casts, whereas dental development, including calcification, alveolar eruption and apical closure can be investigated by means of radiography over the entire period from infancy to maturity.

Eruption of the teeth has been the most frequently used criterion for dental (maturity) age and norms for different populations and both sexes have been established. Different criteria have been applied for tooth eruption, from gingival emergence or the very first appearance of the crown or part of it through the gums to occlusal contact with the opposing teeth.

If gingival emergence is used as a criterion for dental age assessment, it can only be applied from about 6 months up to the age of 30 months for the deciduous dentition, while the eruption of the permanent dentition provides a criterion covering the age 6 to adult age.

Extensively used standards of eruption are the probit transformation based norms of CLEMENTS *et al.* [1957] and KIHLBERG and KOSKI [1954], or the inspectional method by HURME [1949].

In a recent study, DEMIRJIAN *et al.* [1973] pointed out that tooth formation is a more valid indicator of dental maturity than dental eruption. In order to develop a new method of assessing dental age, these authors using panoramic radiographs rated each tooth of 2,928 French Canadian boys and girls aged 2–20 years, according to developmental criteria, such as amount of dentinal deposit and shape change of the pulp chamber. Eight stages of tooth formation, labeled A to H, have been defined and the Tanner-Whitehouse-Healy method for skeletal age assessment has been applied, giving to each tooth a score depending on its stage of development. The scores on all the teeth are then added to give a total maturity score, which can be converted into a dental age using an appropriate table of standards. In case of missing teeth, it is possible to use combinations, but it is not quite satisfactory.

Both methods for dental age assessment (visual eruption age and radiographic formation age) present several drawbacks for non-clinical practical use. Whereas other biological age indicators show that girls are more advanced than boys within chronological age groups, it has been shown [DOERNIG and ALLEN, 1942; ROBINOW *et al.*, 1942; MERIDITH, 1946; ROCHE *et al.*, 1964] that sex differences in deciduous tooth eruption are not significant. Furthermore, deciduous tooth eruption has not been found to be related closely to either stature or skeletal age. Assessment of sex differences in the appearance of permanent teeth is further complicated by the fact that these sex differences vary among teeth [HURME, 1949].

The date of tooth eruption can be recorded only in longitudinal studies, but all the observer can say is that a given tooth erupted at some time between two consecutive examinations. TANNER [1962] suggests to take the age at second examination less half the time elapsed since the first examination to obtain the best estimate of age at eruption. It is obvious that accuracy will be higher the more frequently examinations have been performed.

Finally, it should be remembered that dental age assessment can be used only over limited age ranges, since nearly all the teeth have calcified or erupted by the time puberty begins.

3. *Morphological Size Age*

Height and weight tables have been widely used in assessing nutritional status of the growing and adult individual, but it is questioned if these measures can be used as biological age indicators.

When applied to the appraisal of maturity the interpretation of height and weight (and eventually other anthropometric measures) may easily lead to confusion. As a matter of fact, growth in any dimension is a function of ultimate size and the rate at which that final size will be attained. Moreover, it has been demonstrated [TANNER, 1962] that rate and size are relatively independent, which means that two individuals with equal potential for adult size may reach this final size at different rates and will have different places on growth charts. We also know that skeletally retarded children, at a given skeletal age, may have attained a higher percentage of their adult stature than advanced children of the same age [BAYLEY, 1943]. Thus, the conventional height and weight attained charts are mostly misleading if they are used to assess a child's maturation progress. However, these charts may be used to compare the curve of a child's growth with the centile curves plotted on the conventional growth charts. In childhood, most children stay approximately at the same centile (or in the same channel, as it is sometimes called), whereas at adolescence this technique for assessing growth fails, because the child departs from the 'channel' that he or she has hitherto been following. TANNER *et al.* [1966] developed new standards with 'individual-type' charts to obviate the effect of the adolescent phase difference. To obtain the curves for the new 'individual type' standards, however, it is necessary to use longitudinal records extending over the adolescent phase. Furthermore, height and weight (and eventually other anthropometric measures) charts should be recent and revised every 10 or 15 years until the secular growth trend stops [TANNER, 1962; HEBBELINCK, 1977] and they should be population-specific [HEBBELINCK and BORMS, 1975a].

The whole problem of considering height or weight as primary indicators of morphological growth becomes still more complicated and difficult to apply if one tries to measure the changes in body proportion which occur during the growth process.

Various proportions or anthropometric ratios have been used by several investigators [TANNER, 1962, p. 74], but no clearcut solution has been given as to distinguish differences in proportion due to maturation from differences in proportion among adults, for body proportions change during growth.

4. *Skeletal or Bone Age*

Presently, the evaluation of skeletal or bone age by means of radiography is probably the most extensively used method to assess biological maturity age. Excellent historical surveys have been published by TANNER [1962], ACHESON [1966], ANDERSEN [1968], and the enormous amount of relevant literature on skeletal maturation will not be reviewed here once more.

Skeletal maturity is judged both on the number of ossification centres present and the stage of development of each. Although surface area of the centre was primarily used as an indication of maturity status, shape change is now considered of greater importance. This is a fact similar to the objection made above with regard to the use of anthropometric measures for assessment of maturation, because also the area of the ossification centers depends too much on size of the individual at a given time.

As a rule, the skeletal age assessment is made by comparing the given radiography with a series of standards. In theory, any part of the body could be used for the determination of skeletal age, but in practice the hand-wrist area is the most convenient, because so many ossification centres are present in a small and easily radiographed area without exposing other, perhaps more sensitive parts, of the body. The best (great number of subjects, both sexes and wide age range) and extensively used standards that have been published for the hand and wrist are those of GREULICH and PYLE [1959] and of Tanner-Whitehouse-Healy or the so-called 'TW2 method' [TANNER *et al.,* 1975]. In the 'atlas' method of GREULICH and PYLE, the given hand-wrist radiography is matched successively with standard X-ray photographs and the age is determined with which ossification of the bones most nearly coincides. Although a skilled observer can reach a high degree of reliability on repeated measurement, each such observer seems to set his own level of reading the X-rays. Another difficulty lies in the fact that a metamorphotic process such as the ossification of a bone is being expressed in time measure such as months and years.

ACHESON [1966] and more recently TANNER *et al.* [1975], therefore, established a series of stages through which each bone passes and attributed a numerical weighed score to each of these standardized stages. Centile standard curves can be used for rating the bone maturity and, except at the extremes of the curves, a respective skeletal age can be assigned.

Several authors [ANDERSEN, 1968; BORMS, 1971; ROCHE *et al.,* 1971] stated that skeletal age assessed by the method of GREULICH and PYLE is on the whole lower than the Tanner-Whitehouse method, which is probably due to a difference between the materials upon which the systems for skeletal age assessment are based.

In instances where precision in the appraisal of skeletal maturity of an individual child is of importance, especially in clinical studies, the bone-specific rating method after Greulich-Pyle or Tanner-Whitehouse has been most widely used, while the Greulich-Pyle method of overall skeletal age assessment of the hand-wrist area may be more readily applied in epidemiological research. However, in both methods, the training of the observers, i.e. the readers of the X-rays, should include not only proper objectivity comparisons with experienced readers, but also self-replicate assessments in order to insure the greatest possible reliability [ROCHE *et al.*, 1975]. Eventually, in case of time and budgettary limitations, a simplified method may be used as described by CLARKE and HAYMAN [1962] using but four bones of the hand.

This technique has been applied to adolescent boys only. Using the following four bones, os capitate, metacarpal III, proximal phalanx III, and middle phalanx III, CLARKE and HAYMAN found that the multiple correlation with the total hand was 0.99. In a comparative study on 11-year-old boys, BORMS *et al.* [1973] rated 107 randomly selected X-rays, using Greulich-Pyle, Clarke-Hayman and Tanner-Whitehouse methods. The correlations between the three ratings done independently ranged from 0.85 to 0.90 and the means were not significantly different. Moreover, the reliability (intraobserver) and objectivity (interobserver) tests for the three methods showed correlation coefficients between 0.95 and 0.99.

Concluding Remarks

We have presented in a bird's eye view a synopsis of the most commonly used methods for assessing biological maturity age. Undoubtedly, future growth studies, particularly longitudinal investigations, will provide more information about the different developmental patterns in man.

The techniques used to rate puberal characters are somehow obsolete and accepted by most researchers at face value without having been carefully scrutinized. As GARN and SHAMIR [1958] pointed out, the standards for rating and assessment of sexual maturation may vary within and among populations and the common definitions of the development of puberal characters is not always adequate. Unequivocally, this shows the need for adjustments to and more refinement of the rating scales.

It is further impossible to give any hard and accurate rulings as to the limits of the methods of scaling for the development of secondary sex characters or tooth eruption, because most of the rating is done on the basis of the presence or absence of certain features, but it says little or nothing about when exactly the observed feature began to appear.

There appears also to be a need for constructing norm scales of the various maturity indicators which are specific for the population studied and updating has to be made every 10–15 years in order to adjust for secular shifts and eventual changes in the composition of the population. A major advantage in updating norm scales relates to the better knowledge of the secular trend in a population as well as to better insight in sex-associated differences in biological maturation of boys and girls.

Of course, the small but still existing risk in repeating X-rays for the assessment of tooth formation or bone ossification should be carefully considered and appropriate measures for improving the technique lowering the exposure should be taken.

Finally, it is beyond doubt that, as for the present state of affairs, skeletal maturity assessment is by far the most valid measure of biological maturity age [cf. TANNER, 1962; ANDERSEN, 1968; ROCHE *et al.*, 1975].

Furthermore, skeletal age can be applied over wider age ranges, i.e. from birth to maturity, than any other measure of biological age. A practical issue of skeletal age assessment is not only related to its multiple use in behavioural and biological research or as a tool in clinical diagnostic work, but also to the prediction of an individual's stature [TANNER *et al.*, 1975; ROCHE *et al.*, 1975].

References

ACHESON, R.M.: Maturation of the skeleton; in FALKNER Human development, chap. 16 (Saunders, Philadelphia 1966).

ANDERSEN, E.: Skeletal maturation of Danish school children in relation to height, sexual development, and social conditions. Acta paediat., Stockh. suppl. 185 (1968).

BAYLEY, N.: Size and body build of adolescents in relation to rate of skeletal maturing. Child Dev. *14:* 47–90 (1943).

BORMS, J.: De ossificatietoestand als kriterium voor het bepalen van de biologische ontwikkelingsstand tijdens de groei. Belg. Arch. Soc. Geneesk. *29:* 229–243 (1971).

BORMS, J.; BROEKHOFF, J., and BEUNEN, G.: Vergelijkende studie tussen drie methoden voor het bepalen van de skeletale leeftijd. Belg. Arch. Soc. Geneesk. *31:* 343–353 (1973).

CLARKE, H.H. and HAYMAN, N.R.: Reduction of bone assessments necessary for skeletal age determination in boys. Res. Q. Am. Ass. Hlth phys. Educ. *33:* 202–207 (1962).

Clements, E. M. B.; Davies-Thomas, E., and Pickett, K. G.: Age at which the deciduous teeth are shed. Br. med. J. *i:* 1508–1510 (1957).

Crampton, C. W.: Anatomical or physiological age versus chronological age. Pedagog. Sem. *15:* 230–237 (1908).

Crampton, C. W.: Physiological age – a fundamental principle. Am. phys. Educ. Rev. *13:* 141–154, 268–283, 345–358 (1908).

Demirjian, A.; Goldstein, H., and Tanner, J. M.: A new system of dental age assessment. Hum. Biol. *45:* 211–227 (1973).

Doernig, C. R. and Allen, M. F.: Data on eruption and caries of the deciduous teeth. Child Dev. *13:* 113–129 (1942).

Falkner, F.: General considerations in human development; in Falkner Human development, chap. 2 (Saunders, Philadelphia 1966).

Garn, S. M. and Shamir, Z.: Methods for research in human growth (Thomas, Springfield 1958).

Greulich, W. W., *et al.*: Somatic and endocrine studies of puberal and adolescent boys. Monogr. Soc. Res. Child Dev. 7 (1942).

Greulich, W. W. and Pyle, I. S.: Radiographic atlas of skeletal development of the hand and wrist (Stanford University Press, Stanford 1959).

Grimm, H.: Grundriss der Konstitutionsbiologie und Anthropometrie; 3. Aufl. (Volk & Gesundheit, Berlin 1966).

Hebbelinck, M.: Biological aspects of development at adolescence; in Hill and Mönks Adolescence and youth in prospect, pp. 148–157 (IPC, Guilford 1977).

Hebbelinck, M. and Borms, J.: Biometrische studie van een reeks lichaamsbouwkenmerken en lichamelijke prestatietests van belgische kinderen uit het lager onderwijs. CBGS Technisch Rapport 5 (Ministerie van Volksgezondheid en van het Gezin, Brussel 1975a).

Hebbelinck, M. and Borms, J.: Puberty characteristics and physical fitness of primary school children, aged 6 to 13 years; in Berenberg Puberty. Biologic and psychosocial components, pp. 224–235 (Stenfert Kroese, Leiden 1975b).

Hebbelinck, M. and Borms, J.: Puberteitskenmerken, lichaamsbouw en lichamelijke prestatie bij kinderen uit het lager onderwijs. Belg. Arch. Soc. Geneesk. *34:* 97–109 (1976).

Hurme, V. O.: Ranges of normalcy in the eruption of permanent teeth. J. Dent. Child. *16:* 11–15 (1949).

Hurme, V. O.: Time and sequence of tooth eruption. Symp. on the Human Dentition in Forensic Medicine. J. forensic Sci. *2:* 377–397 (1957).

Kihlberg, J. and Koski, K.: On the properties of the tooth eruption curve. Finska Tandläk. Sällak. Forh. *50:* 6–10 (1954).

Krogman, W. M.: The physical growth of children: an appraisal of studies 1950–1955. Monogr. Soc. Res. Child Dev. *20:* 1956 (1955).

Meridith, H. V.: Order and age of eruption for the deciduous dentition. J. dent. Res. *25:* 43–66 (1946).

Prader, A.: Testicular size. Assessment and clinical importance. Triangle (Sandoz J. med. Sci.) *7:* 240–243 (1966).

Reynolds, E. L. and Wines, J. V.: Physical changes associated with adolescence in boys. Am. J. Dis. Child. *82:* 529–547 (1951).

ROBINOW, M.; RICHARDS, T.W., and ANDERSON, M.: The eruption of deciduous teeth. Growth *6:* 127–133 (1942).

ROCHE, A.F.; BARKLA, D.H., and MARITZ, J.S.: Deciduous eruption in Melbourne children. Austr. dent. J. *9:* 106–108 (1964).

ROCHE, A.F.; DAVILA, G.H., and EYMAN, S.L.: A comparison between Greulich-Pyle and Tanner-Whitehouse assessments of skeletal maturity. Radiology *98:* 273–280 (1971).

ROCHE, A.F., *et al.*: Skeletal maturity of children 6–11 years. US Department HEW, National Health Survey Series II, No. 140, DHEW Publ. No. (HRA) 75–1622.

ROCHE, A.F., *et al.*: Skeletal maturity of youths 12–17 years. US Department HEW, National Health Survey Series II, No. 160, DHEW Publ. No. (HRA) 77–1648.

ROCHE, A.F.; WAINER, H., and THISSEN, D.: Predicting adult stature for individuals. Monogr. Paediat., vol. 3 (Karger, Basel 1975).

SCHWIDETZKY, I.: Eine Typenformel für die Reifungsstufen. Z. menschl. Vererb. Konstit-Lehre *30:* 86–90 (1950).

STOTT, L.H.: Child development. An individual longitudinal approach (Holt, London 1970).

TANNER, J.M.: Growth at adolescence; 2nd ed. (Blackwell, Oxford 1962).

TANNER, J.M., *et al.*: Assessment of skeletal maturity and prediction of adult height (TW 2 method) (Academic Press, London 1975).

TANNER, J.M.; WHITEHOUSE, R.H., and TAKAISHI, M.: Standards from birth to maturity for height, weight, height velocity, and weight velocity: British children, 1965. Parts I and II. Archs Dis. Childh. *41:* 454–471, 613–635 (1966).

WIERINGEN, J.C. VAN, *et al.*: Growth diagrams 1965 Netherlands (Wolters-Noordhoff, Groningen 1971).

WIJN, J.F. DE: Estimation of age at menarche in a population. Maandschr. Kindergeneesk. *33:* 245–252 (1965).

ZELLER, W.: Entwicklungsdiagnose im Jugendalter (Leipzig 1938).

Dr. M. HEBBELINCK, Vrije Universiteit Brussel, *Brussel* (Belgium)

Medicine Sport, vol. 11, pp. 118–123 (Karger, Basel 1978)

Age of Menarche and Motor Performance in Girls Aged 11 Through 18

G. BEUNEN, G. DE BEUL, M. OSTYN, R. RENSON, J. SIMONS and D. VAN GERVEN

Departement Lichamelijke Opvoeding, Katholieke Universiteit Leuven, Heverlee

Age at menarche is claimed to be a useful criterion of the maturation process in girls [JOHNSTON, 1974], despite the fact that it does not indicate a functional reproductive system, and that it does not signal the onset of the growth spurt. Indeed, the major portion of the adolescent growth spurt has occurred by the time of initiation of the menstrual cycle. Age of menarche is also correlated with most of the other processes of adolescence [NICOLSON and HANLEY, 1953; ANDERSEN, 1968; MARSHALL and TANNER, 1969; ONAT and ERBEM, 1974]. It seems, therefore, of interest to analyze the relationships between motor performance and this parameter. These relationships were investigated, during the last 3 months of 1974, as part of a pilot study which examined the physical fitness of Belgian school girls aged 11 through 18.

Materials and Methods

The main purpose of the study was to select valid tests that measure motor performances of girls at these age levels. The test battery used in this study was described by SIMONS *et al.* [1976].

The interrelationships among the motor performance tests, age at menarche and socio-cultural background were also assessed. The tests that were utilized were 24 motor tests, body height, and body weight. The socio-cultural background of the girls, their sports participation, and the age of menarche were also determined through an interview. These interviews were conducted by female interviewers and the girls were asked the following questions related to age of menarche. (1) Do you know what menstruation means? (2) Have you already menstruated? (3) Can you remember the exact date of your first menstruation?

Data were collected on 398 girls from a technical school in Leuven. The total sample was divided into three age groups, i.e. 11–13, 14–15, and 16–18 years. Each age group was divided into maturity categories according to the age at menarche.

Table I. Developmental groups according to age at menarche

Chronological age	Developmental groups
11–13 years	pre menarche post menarche
14–15 years 16–18 years	late: menarche after 13.6 years average: menarche from 12.0 to 13.5 years early: menarche before 11.9 years

In table I, the different maturity categories are given. The first group was subdivided into premenarcheal and postmenarcheal age periods. The two other categories were divided into late, average, and early groups. The reference points for the classification into these groups were chosen on the basis of the date at which the girls reached menarche.

For every group, the statistical characteristics for all dependent variables were calculated. The differences between means were tested for their significance with Student's *t*-test[1].

Results

Before analyzing the differences between the maturity categories, it is of interest to note that the ages of the groups within categories did not differ significantly, as can be seen in table II. In this table, the mean age at menarche in each category is also reported.

For the entire group, the mean age at menarche was 12.8 years (SD = 1.2 years, SE mean = 0.06 years). In tables III and IV, the significant differences that were found between the maturity categories in each age group are indicated. For girls between 11 and 13 years, postmenarcheal girls were taller and heavier and obtained better results for static strength and equilibrium.

Only slight differences occurred between the three maturity groups of 14- through 15-year-old girls. These differences were not significant except for one static strength test (handgrip) and one trunk strength test (leg raiser).

A totally different trend was found for girls 16–18 years. For trunk strength, explosive strength, functional strength, running speed, and speed of limb movement, several highly significant differences were found between

[1] According to DUNCAN [1955], an appropriate level of significance was chosen when more than two groups were involved, to obtain an overall significance level of 1 or 5%.

Table II. Mean ±1 SD for chronological age and age at menarche in different developmental groups

Group	n	Chronological age, years	Age at menarche, years
11–13 years			
Pre	39	12.7 ± 0.55	
Post	29	12.9 ± 0.49	12.0 ± 0.79
14–15 years			
Early	34	14.6 ± 0.61	11.2 ± 0.60
Average	67	14.6 ± 0.60	12.8 ± 0.44
Late	26	14.5 ± 0.51	14.0 ± 0.38
16–18 years			
Early	45	16.7 ± 0.90	11.4 ± 0.54
Average	101	16.9 ± 0.86	12.8 ± 0.44
Late	57	17.0 ± 0.87	14.4 ± 0.67

Table III. Significant differences (*t*-test) between girls 12–15 years of different men archeal age for height, weight and motor fitness components

Component	11–13 years pre-post	14–15 years		
		late-aver.	late-early	aver.-early
Body measures	height 1%	–	–	–
	weight 1%	–	–	–
Static strength	Arm pull 1%	–	–	–
	handgrip 5%	–	handgrip 5%	–
Equilibrium	balance test 5%	–	–	–
Trunk strength	–	–	–	leg raiser 5%
Explosive strength	–	–	–	–
Functional strength	–	–	–	–
Running speed	–	–	–	–
Speed of limb movement	–	–	–	–
Flexibility	–	–	–	–

The level of significance is indicated. All differences are in favor of the *more* mature girls.

Table IV. Significant differences (*t*-test) between girls 16–18 years of different menarcheal age for height, weight and motor fitness components

Component	Late-aver.	Late-early	Aver.-early
Body measurements	–	–	–
Static strength	–	–	
Equilibrium	–	–	–
Trunk strength	sit up 1%	–	–
Explosive strength	stand. broad jump 5%	–	–
Functional strength	–	bent arm hang 5%	–
Running speed	dodge run 1%	–	–
	shuttle run 50 m 1%	shuttle run 50 m 5%	–
	shuttle run 40 m 1%	–	–
	figure 8 duck 1%	–	–
Speed of limb movement	1 foot tapping 1%	1 foot tapping 1%	–
Flexibility	–	–	–

The level of significance is indicated. All differences are in favor of the *less* mature girls.

girls with a late menarche and girls with an early and/or average date of first menstruation. The late maturing girls always obtained better results than the average or early maturing girls.

Discussion

The observed mean age at menarche falls between the boundaries of the ages at menarche previously reported by other investigators [Marshall and Tanner, 1969; Andersen, 1968; Malina *et al.*, 1973] for Caucasian girls. In our opinion, however, the mean age at menarche reported seems rather low, considering the fact that they were from a low socio-economic background. Schmidt [1965] and Renson [1975] have found that while differences in maturation levels are not marked for different socio-cultural groups in Western European countries, they still exist. Wachholder and Cantraine [1976] report a mean age at menarche of 12.9 years for 56 Belgian girls followed longitudinally.

The observed differences between pre- and post-menarcheal 11- to 13-year-old girls are in the same direction as the maturity-performance relationships for boys [Beunen *et al.*, 1974]. However, with increasing age these relationships are inverted for several motor components, resulting in a better performance for late maturing girls at ages 16–18 years.

This inverted relationship is significant for speed tests and tests of explosive, functional and trunk strength, and seems to be in agreement with the findings of other investigators who found that 'the maturity-performance relationships for females, except perhaps for swimmers, is in an opposite direction to that noted for males' [Malina *et al.*, 1973].

Although it is not possible to explain this trend using data from our own investigation, several other authors have attempted to explain this fact. Espenshade and Eckert [1967] and Asmussen [1974] noted that the stagnation or even the decline in motor performance in adolescent girls can be partly explained by a lack of interest in such tasks. In support of this position, Espenshade and Eckert stated: 'It would appear then, that the advent of the menarche marks the peak of steady increase in motor performance in girls but does not necessarily signal the end of all growth'. For the late maturing girls, it is hypothesized that the advent of the menarche occurs later and they possibly were more interested in performing physical tasks during a longer period.

A second explanation that has been advanced is that late maturating girls were apt to develop their motor abilities [Beunen *et al.*, 1976] during a longer participation, resulting in a better performance. Because of better performances, the interest in physical activities was increased.

Finally, it is considered that the late maturing girls are of a more slender body build, or of a more ectomorphic type. The relationship between physique type or somatotype and motor performance in girls has not yet been thoroughly investigated, although the studies of Parizkova [1973] concerning the relationship between body composition and exercise indicate better fitness of non-obese children.

References

Andersen, E.: Skeletal maturation of Danish schoolchildren in relation to height, sexual development and social conditions. Acta paediat. scand., suppl. 185 (1968).

Asmussen, E.: Development patterns in physical performance capacity; in Larson Fitness health and work capacity: international standards for assessment (McMillan, New York 1974).

Beunen, G., *et al.*: Skeletal maturity and physical fitness in 12 to 15 year old boys; in Borms and Hebbelinck Children and exercise. Proc. 5th Int. Symp. Acta paediat. belg. *28:* suppl. 28, pp. 221–232 (1974).

Beunen, G., *et al.*: Motor performance as related to age and maturation. Proc. Int. Committee on Physical Fitness Research, Trois-Rivières 1976.

Duncan, D.B.: Multiple range and multiple F tests. Biometrics 11: 1–45 (1955).

Espenshade, A. and Eckert, H.: Motor performance (Merrill, Columbus 1967).
Johnston, F.E.: Control of age at menarche. Hum. Biol. *46:* 159–171 (1974).
Malina, R.M., *et al.*: Age at menarche in athletes and non-athletes. Med. Sci. Sports *5:* 11–13 (1973).
Marshall, W.A. and Tanner, J.M.: Variations in the pattern of pubertal changes in girls. Archs Dis. Childh. *44:* 291–303 (1969).
Nicolson, A. and Hanley, C.: Indices of physiological maturity deviation and interrelationship. Child Dev. *24:* 3–38 (1953).
Onat, T. and Ertem, B.: Adolescent female height velocity: relationships to body measurement, sexual and skeletal maturity. Hum. Biol. *46:* 199–217 (1974).
Parizkova, J.: Body composition and exercise during growth and development; in Rarick Physical activity, human growth and development (Academic Press, New York 1973).
Renson, R.: Sociocultural determinants of the physical fitness of 13 year old Belgian boys. English issue of the Journal 'Hermes, Leuven', vol.10, pp.349–369 (1975).
Schmidt, M.: Somatische und psychische Faktoren der Reifeentwicklung (Barth, München 1965).
Simons, J., *et al.*: Factor analytic study of motor ability of Belgian girls age 12 to age 19. Int. Congr. of Sport Sciences, Quebec 1976.
Wachholder, A. et Cantraine, F.: Etude de la puberté chez les filles. Association de paramètres somatiques et sociaux. Conf. Etudes de la Croissance de l'Enfant, Brussels 1976.

Dr. G. Beunen, Department Lichamelijke Opvoeding, Katholieke Universiteit Leuven, *Heverlee* (Belgium)

Medicine Sport, vol. 11, pp. 124–133 (Karger, Basel 1978)

Radiographic Age in the Interpretation of Physiological and Anthropological Data

ROY J. SHEPHARD, H. LAVALLÉE, K. M. RAJIC, J. C. JÉQUIER, G. BRISSON and C. BEAUCAGE[1]

Health Sciences Research Centre, University of Québec in Trois Rivières, and Department of Preventive Medicine and Biostatistics, University of Toronto, Toronto, Ont.

Introduction

Radiographic assessments of bone age have been used as an index of maturation since the classical study of BALDWIN *et al.* [1928], and a number of subsequent authors have shown correlations with physical [SHEPHARD, 1971; SZABO *et al.*, 1972; CUMMING *et al.*, 1972] and psychological [HAAS *et al.*, 1971] variables, or have attempted to predict adult stature from bone age and childhood measurements of height [BAYLEY and PINNEAU, 1952; BAYER and BAYLEY, 1976; ROCHE *et al.*, 1975]. The present paper seeks to clarify the contribution that a determination of skeletal age can make to the description of physiological and anthropological data in the pre-pubertal period.

Methods

Subjects. Our subjects were 770 healthy French-Canadian children from the Trois-Rivières region (340 girls and 430 boys, ranging in age from 50 to 150 months). Maturation in this part of Canada occurs relatively late, and almost without exception the students were pre-pubertal. They were chosen randomly from a longitudinal study of growth and development, and included roughly equal numbers from (1) urban and rural areas, and (2) a control group (normal habitual activity) and an experimental group (who had received for 2–3 years an hour of additional physical activity 5 days per week). The parentage of the children was uniformly French Canadian, and the majority of the families had lived in the region for many generations. None of the children had any overt endocrine disorder.

[1] We are much indebted to Drs. R. DUMOULIN, R. MASSIE, S. JÉQUIER and J. LUSSIER for the radiographic examinations, and to M. R. LABARRE for computing services.

Skeletal age. Radiographs of the left wrist (radius, ulna, carpals, metacarpals and phalanges) were classified according to the technique described by GREULICH and PYLE [1959]; these assessments are described hereafter as being based on 'wrist' radiographs. Radiographs of tooth development were graded by the method of GARN *et al.* [1956, 1958, 1965, 1967], which takes account of root closure and crown calcification of the two premolars and two of the three molars in the left mandible; these assessments are described hereafter as based on 'mandibular' radiographs.

Physiological and anthropological variables. Multiple correlation techniques were used to examine relationships between skeletal age and standing height, body weight, Σ muscle force, physical fitness index, Σ skinfold thickness, vital capacity, cardiac volume, aerobic power, and the CAHPER [1966] performance test score (mean of percentiles for six tests). Details of physiological and anthropometric techniques are given elsewhere [SHEPHARD *et al.*, 1968, 1975; WEINER and LOURIE, 1969].

Results

1. Relationship of Bone Age to Chronological Age

'Wrist' age under-estimated chronological age by 10.0 ± 11.9 months ($p < 0.001$), the radiographic assessment being highly correlated with the chronological age ($r = 0.878$ for 699 observations). 'Mandibular' age over-estimated chronological age by 2.6 ± 9.5 months ($p < 0.001$), with a coefficient of correlation of 0.909 between the two variables.

The discrepancy between 'wrist' age and skeletal age (fig. 1) remained

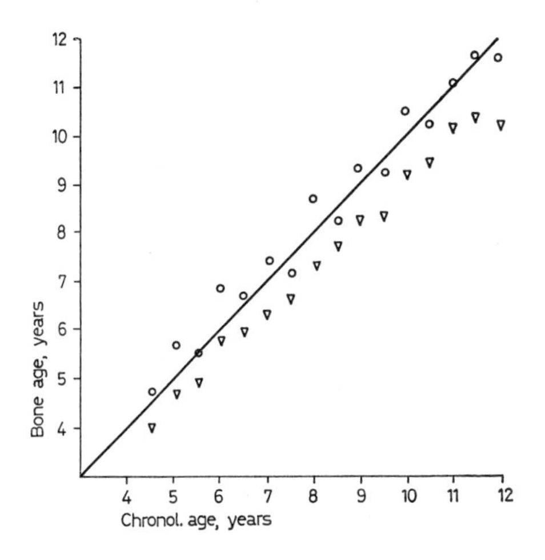

Fig. 1. The relationship between chronological age and radiographic age. ▽ = Observations on 699 wrist radiographs; ○ = observations on 677 mandibular radiographs.

rather constant over the age range studied, although there was a tendency to an increase in the variance of the discrepancy as puberty was approached. On the other hand, 'mandibular' age over-estimated chronological age in the younger children, and under-estimated it in those who were older.

2. *Factors Modifying Skeletal Maturity*

The influence of sex, milieu, and activity was examined by fitting linear regressions relating skeletal age to chronological age within each sub-group of the data. For purposes of comparison, all results have been interpolated to a common chronological age of 100 months (table I). Since there is some evidence that the effect of our experimental variables may be age-specific, the data have also been subjected to graphic analysis (fig. 2), and finally interaction between factors has been checked by a form of multiple regression analysis.

Table I. Factors modifying the discrepancy between chronological and radiographic age. Linear regression equations, bone ages at chronological age of 100 months, and coefficients of correlation between chronological and radiographic ages. Results for observer A (770 films) and observer D (283 wrist and 261 mandibular films)

Observer and subject group	Wrist radiograph			Mandibular radiograph		
	equation	bone age$_{100}$	r	equation	bone age$_{100}$	r
Boys A	−5.9+0.94 (A)	88.2	0.891	7.1+0.94 (A)	101.6	0.923
D	−6.8+0.99 (A)	92.1	0.938	2.2+0.94 (A)	96.7	0.939
Girls A	10.8+0.81 (A)	91.7	0.869	12.7+0.90 (A)	103.5	0.886
D	0.9+0.93 (A)	94.1	0.897	6.9−0.90 (A)	97.1	0.932
Δ Boys/girls A		−3.5±1.1			−1.9±0.8	
D		−2.0±1.3			−0.4±1.1	
Urban A	−1.1+0.90 (A)	89.2	0.892	2.2±0.98 (A)	100.5	0.937
Rural A	3.7+0.86 (A)	89.4	0.864	25.3+0.79 (A)	104.3	0.852
Δ Urban/rural A		−0.2±0.9			−3.8±0.7	
Exercised A	8.3+0.81 (A)	89.6	0.593	−9.4+1.15 (A)	105.3	0.819
D	1.5+0.90 (A)	90.9	0.584	2.1+0.95 (A)	97.2	0.720
Control A	−2.1+0.91 (A)	88.8	0.899	11.5+0.91 (A)	102.0	0.916
D	−4.3+0.97 (A)	93.3	0.942	5.4+0.92 (A)	97.4	0.941
Δ Exercise/control A		+0.8±1.4			+3.3±1.0	
D		−2.4±1.0			+0.4±0.9	

(a) Sex. The 'wrists' of the girls were less retarded than those of the boys relative to the Greulich-Pyle norms ($p < 0.001$). This difference was due mostly to girls 8 years of age and younger (fig. 2). 'Mandibular' development showed a smaller but parallel sex difference ($p < 0.02$).

(b) Urban/rural differences. Mean 'wrist' ages showed no urban/rural differences at a chronological age of 100 months. However, regressions for the rural milieu were characterized by a shallow slope and a large zero intercept, so that the youngest rural girls had more advanced 'wrist' development that their urban counterparts (fig. 2). The mandibular radiographs showed a similar trend. At 100 months, the rural children had an advantage of 3.8 ± 0.7 months ($p < 0.001$), and graphic analysis showed that the discrepancy was due to children under 9 years of age, both boys and girls (fig. 2).

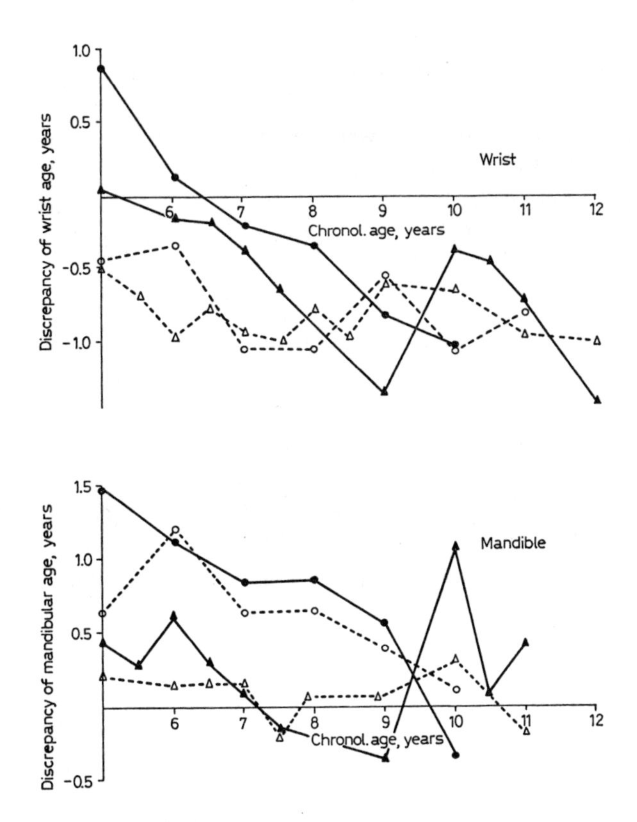

Fig. 2. Discrepancy between chronological and radiographic age. Results classified for sex, milieu and chronological age. ● = Rural girls; ○ = rural boys; ▲ = urban girls; △ = urban boys.

(c) Added physical education. The effects of added physical activity were inconsistent. Wrist radiographs showed no significant differences between the two groups, but mandibular development was significantly more advanced in the exercised children.

(d) Multiple regression analysis. Possible interactions of sex, milieu. and added exercise were investigated by fitting multiple regressions of the type:

$$\text{Age} = \text{A (sex)} + \text{B (milieu)} + \text{C (activity)} + \text{D}.$$

For the purpose of this analysis, females, a rural environment, and added daily exercise were scaled as 1.0, while males, an urban environment, and control activity patterns were scaled as 0. The equations developed were as follows:

$$\Delta \text{ age ('wrist')} = 4.56 \text{ (sex)} + 0.39 \text{ (milieu)} - 3.59 \text{ (activity)} - 13.0,$$
$$\Delta \text{ age ('mandible')} = 1.33 \text{ (sex)} + 4.53 \text{ (milieu)} + 1.47 \text{ (activity)} - 0.28.$$

Significant effects on 'wrist' maturation were demonstrated for sex ($F_{1,614}$= 23.6) and added exercise ($F_{1,614}$= 13.5). In the case of the 'mandible', maturation was advanced in the rural milieu ($F_{1,614}$= 33.9); effects of added physical activity ($F_{1,614}$= 3.4) and sex ($F_{1,614}$= 3.1) did not reach statistical significance.

3. Relationship of Skeletal Maturity to Physiological and Anthropological Data

The contribution of skeletal age to the description of selected physiological and anthropological variables is examined by sequential multiple regression (table II). At the first stage, effects due to age and $(\text{age})^2$ are indicated. Partial correlations due to height and weight are next added to the cumulative variance (Σr^2), since skeletal age may be serving largely as an index of these two readily measured variables. The final columns indicate the residual contributions (Δr^2) of 'wrist' age, 'mandibular' age, and 'wrist' plus 'mandibular' age to a description of the total variance in the selected measurements.

The 'wrist' radiographs generally contributed a larger Δr^2 than the 'mandibular' radiographs, although because of limitations in the possible accuracy of the assessments, and perhaps more importantly the close correlations of skeletal and chronological age in the pre-pubertal child, the performance of neither radiographic assessment was particularly impressive. The largest effects of skeletal age were on height (5.3% of total variance), weight

Table II. Contributions of radiographic age to the description of physiological and anthropological data. (Results expressed as Δr^2, after allowance for chronological age, age^2, and where appropriate height and weight)

Variable	Σr^2 age+age^2	Σr^2 age+age^2 +height +weight	Δr^2		
			wrist	man-dible	wrist and mandible
Height	0.756	–	0.052	0.008	0.053
Weight	0.370	–	0.080	0.027	0.094
Σ Muscle force	0.374	0.528	0.015	0.004	0.016
Physical fitness index	0.027	0.213	0.076	0.001	0.076
Σ Skinfold	0.034	0.606	0.001	0.000	0.000
Vital capacity	0.546	0.753	0.002	0.000	0.002
Cardiac volume	0.418	0.643	0.002	0.004	0.007
$\dot{V}o_2$ (max)/kg	0.022	0.048	0.000	0.009	0.010
Σ CAHPER tests (percentile)	0.037	0.122	0.036	0.013	0.042

Table III. Correlations between selected variables and the discrepancy of radiographic and chronological ages. Observations of radiographs by Dr. A., data expressed as r^2 and F ratios, the latter having 1 and 173–363 degrees of freedom for different variables[1]

Variable	Wrist		Mandible	
	r^2	F ratio	r^2	F ratio
Height	0.004	n.s.	0.024	8.60
Arm span	0.008	n.s.	0.022	8.39
Weight	0.002	n.s.	0.006	n.s.
ΣMuscle force	0.036	6.60	0.006	n.s.
Σ Muscle force/kg	0.008	n.s.	0.005	n.s.
Physical fitness index	0.003	n.s.	0.007	n.s.
PFI/kg	0.008	n.s.	0.012	n.s.
Σ Skinfold	0.024	8.72	0.008	n.s.
Σ Skinfold/kg	0.007	n.s.	0.000	n.s.
Vital capacity	0.004	n.s.	0.012	4.36
Cardiac volume	0.001	n.s.	0.001	n.s.
Cardiac volume/kg	0.013	4.55	0.005	n.s.
$\dot{V}o_2$ (max)/kg	0.002	n.s.	0.020	4.98
Σ CAHPER tests (percentiles)	0.018	5.89	0.008	n.s.

[1] After allowance for height and weight, Δr^2 for the radiographs was generally diminished, exceptions being the physical fitness index (0.051 and 0.000) and Σ CAHPER percentiles (0.056 and 0.019).

(9.4%), physical fitness index (7.6%) and percentile score on the CAHPER performance battery (4.2%). Rather larger apparent effects would have emerged if we had neglected to introduce height and weight as independent variables at a preliminary stage in the analysis. Thus, allowing simply for chronological age and $(\text{age})^2$, the two radiographs would add 8.4% to the description of Σ muscle force, 6.2% to Σ skinfolds, 3.6% to vital capacity. 4.8% to cardiac volume, and 1.5% to $\dot{V}O_2$ (max)/kg; only the CAHPER test (1.4%) would suffer from the neglect of height and weight.

A second method of analysis is to plot the difference between radiographic and chronological age against the physiological variables (table III). Even before allowing for height and weight, the correlations were in most instances extremely feeble, the best description (Σ muscle force) accounting for no more than 3.6% of the variance in wrist age. Whether corrected by the use of simple ratios or through formal multiple regression analysis, the general effect of introducing height and/or weight was to reduce the usefulness of the radiographs; two notable exceptions were the physical force index (partial correlation after allowance for height and weight 5.1% of wrist radiograph variance) and the Σ CAHPER test (partial correlation 5.6% of the 'wrist' radiograph and 1.9% of the 'mandibular' radiograph variance).

Discussion

1. Differences from published norms. The disparate development of wrist and mandible relative to published norms is perhaps to be anticipated from GARN's [1956, 1958, 1959, 1965] studies of communality. Nevertheless, the differences observed in our study could reflect in part differences of maturation between the standard populations of GREULICH and PYLE and GARN *et al.*, rather than an unherent difference in the maturation of the wrist and mandible.

The children of the Cleveland region examined by GREULICH and PYLE [1959] were somewhat privileged from an economic standpoint [MICHAUT *et al.,* 1972]. Studies from Melbourne [ROCHE *et al.*, 1971], Berkeley [BAYLEY and PINNEAU, 1952] and Toronto [SHEPHARD *et al.*, 1968] have shown comparable rates of maturation, but slower development has been described in 'faculty' children in Nebraska [FRY, 1971], malnourished rural children in the US [ABBOTT *et al.*, 1950], city-dwellers born of immigrant parents [JOHNSTON and JAHINA, 1965] and children from various under-developed communities [BERENI *et al.*, 1971; BLANCO *et al.*, 1972; LEVINE, 1972; MACKAY, 1961]. Some retardation of maturation was anticipated in the French-Canadian children, since previous studies had demonstrated a short stature relative to children of the same age living in Toronto [SHEPHARD *et al.*, 1975].

The reasons for earlier maturation of economically privileged groups are still discussed. In under-developed countries and some rural areas of the United States, poor nutrition may retard development [MAZESS and CAMERON, 1972]. However, there was little evidence of malnutrition in the present sample of francophone children. Climatic factors [CLEGG *et al.*, 1972] and disease [FRANCIS, 1940] can also be excluded. A more important consideration is likely genetic. Both maturation and adult stature seem enhanced by genetic admixture, and whereas Toronto and the major US cities are very cosmopolitan, the Trois Rivières region is populated almost exclusively by the descendants of a small group of early settlers from France.

Irrespective of mechanisms, there seemed no retardation of tooth development relative to children of Yellow Springs, Ohio [GARN *et al.*, 1967]. Nor was the difference in 'wrist' maturation enhanced by living in the rural areas around Trois Rivières; indeed, tooth development was further advanced in our rural than in our urban sample. A negative effect of exercise upon bone development has been described previously [SUZUKI, 1970]. Our data confirmed this for the wrist (where there was presumably direct exposure to mechanical effects of activity). However, the mandibular radiographs suggested an enhanced maturation in the more active children; here, there may have been a response to hormonal factors unimpeded by secondary mechanical influences.

2. *Description of physiological variables*. Previous reports have shown that skeletal age can contribute to the description of anthropological data [SZABO *et al.*, 1972], aerobic power [SHEPHARD, 1971] and track performance [CUMMING *et al.*, 1972]. In all three studies, the children were older than the present group. SZABO *et al.* [1972] and SHEPHARD [1971] eliminated age by studying homogenous groups; SZABO *et al.* [1972] made no allowance for possible effects of height and weight, but SHEPHARD [1971] pointed out that the correlation of aerobic power with radiographic age became insignificant when the former was standardized for body weight. CUMMING *et al.* [1972] claimed that in 11- to 18-year-old children, bone age was a better predictor of track and field performance than chronological age, even after allowance for the effects of height and weight; however, the published material did not show how much bone age added to a description based on chronological age, height and weight.

The present results show that in young children the determination of radiographic age adds little to the description of physiological variables yielded by age, $(age)^2$, height and weight. Further, this seems an inevitable consequence of (1) a substantial measurement error, and (2) a very strong correlation between chronological and radiographic age. In our series, about 85% of the variance in radiographic age was attributable to chronological age. Of the remaining 15% variance, perhaps 4% was attributable to measurement errors, leaving at most 11% new information to be used in the description of anthropological and physiological data. Much of the final 11% is in turn taken up by partial correlations with height and weight, and even in such tests as the CAHPER performance battery and the physical fitness index, radiographic age has little residual predictive value.

References

ABBOTT, O.D.; TOWNSEND, R.O.; FRENCH, R.B., and ASHMANN, C.F.: Carpal and epiphysial development. Another index of nutritional status of rural school children. Am. J. Dis. Child. *19:* 69–81 (1950).

BALDWIN, B.T.; BUSBY, L.M., and GARSIDE, H.V.: Anatomic growth of children. A study of some bones of the hand, wrist and lower forearm by means of roentgenograms. University of Iowa Study of Child Welfare, vol. 4, p. 88 (1928); cited by CUMMING *et al.* (1972).

BAYER, L.M. and BAYLEY, N.: Growth diagnosis; 2nd ed. (University of Chicago Press, Chicago 1959).

BAYLEY, N. and PINNEAU, S.R.: Tables for predicting adult height from skeletal age: Revised for use with the Greulich-Pyle hand standards. J. Pediat. *40:* 423–441 (1952).

BERENI, L.; HECHES, P.; BLANCHOT, M. et DUVINAGE, J.F.: Enquête préliminaire sur la croissance osseuse en milieu scolaire à Saint Louis. Bull. Soc. Med. Afr. Noire Igue Frse *16:* 559–563 (1971).

BLANCO, R.A.; ACHESON, R.M.; CANOSA, C., and SALOMON, J.B.: Sex differences in retardation of skeletal development in rural Guatemala. Pediatrics *50:* 912–915 (1972).

CAHPER (Canadian Association for Health, Physical Education and Recreation): The CAHPER Fitness-Performance Test Manual. For boys and girls 7 to 17 years of age (Canadian Association for Health, Physical Education and Recreation, Toronto 1966).

CLEGG, E.J.; PAWSON, I.G.; ASHTON, E.H., and FLINN, R.M.: The growth of children at different altitudes in Ethiopia. Phil. Trans. R. Soc. *264:* 403–437 (1972).

CUMMING, G.R.; GARAND, T., and BORYSYK, L.: Correlation of performance in track and field events with bone age. J. Pediat. *80:* 970–973 (1972).

FRANCIS, C.C.: Factors influencing appearance of centers of ossification during early childhood. Am. J. Dis. Child. *59:* 1006–1012 (1940).

FRY, E.I.: Tanner-Whitehouse and Greulich-Pyle skeletal age velocity comparisons. Am. J. Phys. Anthrop. *35:* 377–380 (1971).

GARN, S.M.; LEWIS, A.B., and POLACHECK, D.L.: Variability of tooth formation. J. dent. Res. *38:* 135–148 (1959).

GARN, S. M.; LEWIS, A. B.; KOSKI, K., and POLACHECK, D. L.: The sex difference in tooth calcification. J. dent. Res. *37:* 561–567 (1958).

GARN, S. M.; LEWIS, A. B., and BLIZZARD, R. M.: Endocrine factors in dental development. J. dent. Res. *44:* 243–258 (1956).

GARN, S. M.; LEWIS, A. B., and KEREWSKY, R. S.: Genetic, nutritional and maturational correlates of dental development. J. dent. Res. *44:* 228–242 (1965).

GARN, S. M.; ROHMAN, C. G., and SILVERMAN, F. N.: Radiographic standards for postnatal ossification and tooth calcification. Med. Radiogr. Photogr. *43:* (1967).

GREULICH, W. W. and PYLE, S. I.: Radiographic atlas of skeletal development of the hand and wrist; 2nd ed. (Oxford University Press, London 1959).

HAAS, J. D.; HUNT, E. E., and BUSKIRK, E. R.: Skeletal development of non-institutionalized children with low intelligence quotients. Am. J. phys. Anthrop. *35:* 455–466 (1971).

JOHNSTON, F. E. and JAHINA, S. B.: Contribution of carpal bones to assessment of skeletal age. Am. J. phys. Anthrop. *23:* 349–354 (1965).

LEVINE, E.: The contributions of the carpal bones and the epiphyseal centres of the hand to the assessment of skeletal maturity. Hum. Biol. *44:* 317–327 (1972).

MACKAY, R. H.: Skeletal maturation. Med. Radiogr. Photogr. *37:* 14–15 (1961).

MAZESS, R. B. and CAMERON, J. R.: Skeletal growth in children: maturation and bone mass. Am. J. phys. Anthrop. *35:* 399–408 (1972).

MICHAUT, E.; NIANG, I. et DAN, V.: La maturation osseuse pendant la période pubertaire. A propos de l'étude de 227 adolescents dakarois. Ann. Radiol. *15:* 767–779 (1972).

ROCHE, A. F.; DAVILA, G. H., and EYMAN, S. L.: A comparison between Greulich-Pyle and Tanner-Whitehouse assessments of skeletal maturity. Radiology *98:* 273–280 (1971).

ROCHE, A. F.; WEINER, X. X., and THISSEN, X. X.: Predicting adult stature for individuals. Monogr. Paediat., vol. 3 (Karger, Basel 1975).

SHEPHARD, R. J.: The working capacity of schoolchildren; in SHEPHARD Frontiers of fitness (Thomas, Springfield 1971).

SHEPHARD, R. J.; ALLEN, C.; BAR-OR, O.; DAVIES, C. T. M.; DEGRÉ, S.; HEDMAN, R.; ISHII, K.; KANEKO, M.; LACOUR, J. R.; DIPRAMPERO, P. E., and SELIGER, V.: The working capacity of Toronto schoolchildren. Can. med. Ass. J. *100:* 560–566 705–714 (1968).

SHEPHARD, R. J.; LAVALLÉE, H.; LARIVIÈRE, G., *et al.:* La capacité physique des écoliers Canadien-français. Une comparaison avec les échantillons Anglo-Canadiens et Esquimaux. Union Méd. *103:* 1767–1777 (1974); *104:* 259–269, 1131–1136 (1975).

SUZUKI, S.: Experimental studies on factors in growth, monogr. 35, pp. 6–11 (Soc. Res. Child Development, Washington 1970).

SZABO, S.; DOKA, J.; APOT, P. und SOMOGYVARI, K.: Die Beziehungen zwischen Knochenlebensalter, funktionellen anthropometrischen Daten und der aeroben Kapazität. Schweiz. Z. Sportmed. *20:* 109–115 (1972).

WEINER, J. S. and LOURIE, J. A.: Human biology – a guide to field methods (Blackwell, Oxford 1969).

Dr. R. J. SHEPHARD, Department of Preventive Medicine and Biostatistics, University of Toronto, *Toronto, Ont.* (Canada)

Medicine Sport, vol. 11, pp. 134–139 (Karger, Basel 1978)

Secular Growth Trend Data in Belgian Populations Since 1840

A Dimensional and Proportional Analysis

A. S. VAJDA and M. HEBBELINCK

Laboratory of Human Biometry and Movement Analysis,
Vrije Universiteit Brussel, Brussels

Introduction

In 1840, the well-known Belgian physical anthropologist QUETELET [1842] carried out a large cross-sectional survey of Belgian children, producing the first set of norms which were used clinically for over 70 years in Europe. Subsequent to QUETELET's surveys, there have been four major cross-sectional studies in Belgium which include mean height and weight values as summarized in table I.

Secular trend data, classically looked upon in absolute terms as discussed by TANNER [1964], VAN WIERINGEN [1972] and corroborated studies elsewhere show an apparent universal secular trend for both increased stature and body weight. The mean values for boys and girls from the Belgian studies show this clearly with nearly all reported mean values in every study since 1840, being larger than previously stated values at each age level as shown in table II.

Since the development of the phantom or unisex human prototype by ROSS and WILSON [1974], it is now possible to proportionally analyze anthropometric data in terms of a human standard. VAJDA *et al.* [1974] showed that boys measured in 1840 at ages 6, 7, 8 and 9 were proportionally heavier for their height than the boys in subsequent studies and at ages 10, 11 and 12 they became proportionally lighter than the modern boys as shown in table III. The girls measured in 1840 were not always heavier than the girls of 1924, 1929/30 and 1960 at ages 6, 7, 8, 9, 10, 11 and 12, but always heavier than the girls measured in the most recent Performance and Talent Study [HEBBELINCK and BORMS, 1975]. Furthermore, the 6-year-old is always heavier for his height as compared to the 12-year-old and the girls who are lighter at age 6 approach male values by age 12 in 1924, 1929/30 and 1960, but in 1840 and 1970 the boys were proportionally lighter.

Consideration of secular trend data in dimensional terms as discussed by ÅSTRAND and RODAHL [1972], and HEBBELINCK and ROSS [1974], may reveal more information about the status of growth over the years. The pur-

Table I. Cross-sectional studies of Belgian children

1840	QUETELET, M. A.: A treatise on man and the development of his faculties, Edinburg, 1842 (n = ?)
1924	Ministry of Art and Science, Committee for Relief of Belgium: Hoover Commission (n = 18,000)
1929/30	Ministry of Public Affairs and Education (n = 17,000)
1960	Ministry of Health and Welfare (n = 55,000 boys, 45,000 girls)
1967/68	Performance and Talent Project (n = 7,500 boys, Netherlandish- and French-speaking)
1971	Performance and Talent Project (n = 5,500 girls, netherlandish-speaking)

Table II. Height and weight of Belgian boys and girls from 1840 until 1971

Years	Sex	1840 Ht, cm	Wt, kg	1924		1929/30		1960		1967/68 M 1971 F	
6	M	104.7	17.2	109.7	18.4	110.3	19.6	114.0	20.5	116.3	20.9
	F	103.1	16.0	109.2	18.1	109.5	19.5	113.5	19.5	116.7	20.5
7	M	110.5	19.1	115.0	20.3	116.3	21.5	119.5	22.5	121.3	23.0
	F	108.6	17.5	114.7	19.9	115.3	20.8	118.8	21.5	120.9	22.2
8	M	116.2	20.8	120.5	22.4	120.9	23.2	125.5	25.0	127.0	25.6
	F	114.1	19.1	119.8	21.7	120.5	22.6	123.5	24.0	126.6	24.6
9	M	121.9	22.7	124.8	24.3	126.3	25.0	130.0	27.5	132.0	28.2
	F	119.5	21.4	124.5	23.8	126.1	24.8	129.0	27.0	131.7	27.1
10	M	127.5	24.5	129.5	26.4	130.2	27.6	135.0	30.0	136.8	31.0
	F	124.8	23.3	130.4	26.5	130.1	26.9	134.0	30.0	136.6	29.9
11	M	133.0	27.1	134.1	28.8	134.8	29.5	140.0	33.0	141.4	33.8
	F	129.9	25.7	134.8	29.0	135.3	29.7	139.5	34.0	142.7	33.9
12	M	138.5	29.8	138.2	31.2	139.4	32.9	144.5	37.0	145.5	36.6
	F	135.3	29.8	140.0	32.2	141.5	34.5	146.0	38.5	147.6	37.3
13	M	143.9	34.4	143.0	34.0	143.8	35.7	149.5	41.0	149.1	39.7
	F	140.3	32.9	145.8	36.2	147.2	38.2	151.5	43.5	150.6	41.4

Note: 1971 girls data from Flemish sample only.

Table III. Proportional z values for body weight for boys and girls based on the phantom stratagem

Age	Sex	1840	1924	1929/30	1960	1967/68 1971
6	M	1.10	0.48	0.86	0.42	0.11
	F	0.86	0.46	0.83	0.13	−0.12
7	M	0.60	0.14	0.32	0.05	−0.12
	F	0.34	0.05	0.27	−0.16	−0.31
8	M	0.07	−0.17	0.01	−0.26	−0.35
	F	−0.15	−0.29	−0.11	−0.21	−0.56
9	M	−0.34	−0.34	−0.40	−0.34	−0.48
	F	−0.34	−0.44	−0.42	−0.30	−0.71
10	M	−0.73	−0.54	−0.34	−0.52	−0.57
	F	−0.63	−0.66	−0.51	−0.36	−0.79
11	M	−0.91	−0.66	−0.61	−0.62	−0.66
	F	−0.80	−0.72	−0.64	−0.33	−0.82
12	M	−1.08	−0.74	−0.53	−0.48	−0.70
	F	−0.61	−0.78	−0.55	−0.42	−0.86
13	M	−0.90	−0.85	−0.63	−0.48	−0.68
	F	−0.67	−0.82	−0.65	−0.31	−0.56

pose of this paper is to investigate the dimensional relationship between height and weight for the Belgian secular data and compare the results to phantom z scores.

Material and Method

The dimensional model is based on the general reasoning that all body length, breadths, girths, skinfold thickness and times have the dimension L, areas such as body surface and cross section of muscles, L^2 and weights and volumes, L^3. Given that shape and composition are constant, then theoretically if height is used as the criterion for size, weight would became dimensionless when it is proportional to h^3.

Exponential values for the mean height and weight values shown in table II were calculated over the age ranges 6–13, 6–11, 6–10, 6–9, 6–8 and then 6–7, 7–8, 8–9, 9–10, 10–11, 11–12 and 12–13 using a double logarithmic equation in the form of equation 1:

$$\text{log weight} = a + b \text{ log height}, \qquad (1)$$

where 'b' equals the exponential value and 'a' a constant.

In this way, proportional changes are not lost when only mean values are available according to VAN WIERINGEN [1972].

Table IV. Exponential values for mean height and weight values for an increasing age range

	1840		1924		1930		1960		1970	
	M	F	M	F	M	F	M	F	M	F
6–7	1.94	1.72	2.08	1.93	1.75	1.65	1.98	2.14	2.27	2.25
6–8	1.83	1.75	2.09	1.94	1.83	1.75	2.07	2.45	2.31	2.24
6–9	1.81	1.95	2.15	2.07	1.81	1.85	2.22	2.57	2.36	2.30
6–10	1.79	1.99	2.18	2.15	2.00	1.97	2.27	2.63	2.42	2.38
6–11	1.85	2.06	2.22	2.24	2.06	2.08	2.33	2.71	2.46	2.48
6–12	1.92	2.23	2.28	2.32	2.18	2.27	2.46	2.74	2.50	2.55
6–13	2.08	2.34	2.32	2.40	2.27	2.38	2.55	2.80	2.56	2.68

Table V. Exponential values for mean height and weight values between two age ranges from age 6 to 13 according to a double logarithmic formula

	1840		1924		1930		1960		1970	
	M	F	M	F	M	F	M	F	M	F
6–7	1.94	1.72	2.08	1.93	1.75	1.65	1.98	2.14	2.27	2.25
7–8	1.70	1.77	2.11	1.95	1.96	1.88	2.15	2.84	2.33	2.23
8–9	1.83	2.46	2.32	2.45	1.71	2.04	2.71	2.70	2.50	2.45
9–10	1.70	1.96	2.24	2.32	3.25	2.60	2.31	2.77	2.65	2.69
10–11	2.39	2.45	2.49	2.47	1.92	2.53	2.62	3.11	2.61	2.87
11–12	2.34	3.63	2.66	2.77	3.25	3.34	3.61	2.73	2.78	2.83
12–13	3.75	2.73	2.52	2.88	2.63	2.58	3.02	3.30	3.33	5.18

Results and Discussion

Exponential values for the age ranges, as discussed, are presented in tables IV and V. The values show how much height is required for the investigated quantity weight and thus the lower the exponent the less height is needed to express a given weight. From table IV, we see that modern children approach the theoretical value $(w) = h^3$ and comes in close agreement with Danish cross-sectional data [ASMUSSEN and HEEBÖLL-NIELSEN, 1955] on subjects 7–16 where a height dimension value of 2.68 for boys and 2.88 for girls with weight was found. The differences in our values and those of ASMUSSEN and HEEBÖLL-NIELSEN may be due to a difference in age range. One sees from table IV that with increasing age range the exponential value increases – perhaps because there are more plottings points.

Table V expresses in fact the velocity of change between two age groups. From these values, we can estimate whether the child is growing in height or weight. An increased exponential values indicates weight increases at a slightly higher rate with increasing height and similarily a decreasing value indicates weight increases are slightly lower to increasing height.

Conceptual clarity is possible when the results are seen in terms of phantom z scores. The dimensional relationship between two absolute values shows the dependence of one value on the other but relativeness of growth patterns are clearly shown when using the phantom model. Dimensionally, the modern child needs more height to express a given weight but proportionally he is seen as sometimes lighter for his height.

Findings from these two approaches evokes further questioning. Why is growth apparently more manifest in stature than weight in present-day children or, for that matter, how valid are body surface area formulae based on height and weight when these two parameters may be disproportional components in the morphological make-up of different subjects who may reflect ethnic, regional or secular differences from causes yet known?

References

ÅSTRAND, P. O. and RODAHL, K.: A textbook of work physiology (McGraw Hill, New York 1970).

ASMUSSEN, E. and HEEBÖLL-NIELSEN, K.: A dimensional analysis of physical performance and growth in boys. J. appl. Physiol. *7:* 593 (1955).

HEBBELINCK, M. H. and BORMS, J.: Biometrische Studie van een reeks Lichaamsbouwkenmerkenen. Lichamelijke Prestatietests van Belgische kinderen uit het Lager Onderwijs. Technisch rapport 5, Centrum voor Bevolkings- en Gezinsstudiën, Ministerie van Volksgezondheid en van het Gezin (1975).

HEBBELINCK, M. and ROSS, W. D.: Body type and performance; in LARSON International Standards for the Assessment of Physical Fitness and Work Capacity (Macmillan, New York 1974).

Ministère de l'Instruction Publique: Considérations sur quelques notions biologiques élémentaires relatives sur la croissance et sur l'opportunité d'une surveillance périodique de la taille et du poids de l'écolier. Valeurs moyennes, évaluées en semestres, de la taille et du poids de l'enfant belge entre 3 et 18 ans (1934).

Ministerie van Kunsten en Wetenschappen: Tabellen van de gestalte en het gewicht der Belgische kinderen van 3 tot 15 jaar (1925).

Ministerie van Volksgezondheid en van het Gezin: Biometrie gegevens (mimeographed document) (1960).

QUETELET, M.A.: A treatise on man and the development of his faculties (Chambers, Edinburg 1842).

ROSS, W.D. and WILSON, N.C.: A stratagem for proportional growth assessment. Acta paediat. belg. *28:* 169–182 (1974).

TANNER, J.M.: The physique of the olympic athlete, pp. 108–109 (Unwin Brothers, Woking 1964).

VAJDA, A.; HEBBELINCK, M., and ROSS, W.D.: Size and proportional weight characteristics of Belgian children since 1840. Proc. 6th Int. Symp. on Paediatric Work Physiology, Seč 1974.

WIERINGEN, J.C. VAN: Seculaire Groeiverschuiving, Lengte en Gewicht Surveys 1964–1966 in Nederland in Historisch Perspectief (Nederlands Instituut voor Preventieve Geneeskunde TNO, Leiden 1972).

A.S. VAJDA, Laboratory of Human Biometry and Movement Analysis, Vrije Universiteit Brussel, *Brussels* (Belgium)

Medicine Sport, vol. 11, pp. 140–151 (Karger, Basel 1978)

Size Dissociation of Maximal Aerobic Power during Growth in Boys[1]

D. A. BAILEY, W. D. ROSS, R. L. MIRWALD and C. WEESE
University of Saskatchewan, Saskatoon, Sask. and Simon Fraser University, Burnaby, B. C.

Introduction

Maximal aerobic power as measured by maximal oxygen uptake ($\dot{V}O_2$ max) is a universally accepted parameter for measuring status and change in cardiovascular fitness. Not so universally accepted is how to express maximal power in relation to body size. In comparing individuals of different body size, the traditional approach has been to relate maximal oxygen uptake to body weight, under the assumption that $\dot{V}O_2$ max is directly proportional to body mass ($\dot{V}O_2 \max \propto M^{1.0}$). However, as noted by ÅSTRAND and RODAHL [1970], there is a paucity of data available on maximal values for oxygen uptake in animals of different size to test this assumption. Further, there are theoretical and practical reasons for questioning this approach.

TENNEY [1967], while stating that the little experimental data available appear to show that $\dot{V}O_2$ max is nearly proportional to $M^{1.0}$, admits that hemodynamically there are good reasons for predicting a direct proportionality between maximal power and resting power. A simple surface: volume relationship would suggest that resting oxygen uptake in animals should be proportional to $M^{2/3}$. Some investigators suggest that shape factors and other considerations bring this proportionality closer to $M^{3/4}$ [KLEIBER, 1961]. There is no evidence that resting power is directly proportional to $M^{1.0}$.

Practically, the question of how maximal aerobic power relates to body mass and size is of considerable significance. In fitness training studies on children there is a problem in determining if changes in maximal aerobic

[1] This study was supported by a National Health Research Grant of Canada No. 608-1051-44 and a National Research Council of Canada Grant No. 711-007.

power are a result of training, growth or both, since increasing size may result in changes similar to the training effect. Recently, several investigators have tried to control for growth changes on the basis of expected change due to increasing linear dimensions, then, variations from the expected were attributed to training effects [ERIKSSON and KOCH, 1972; ANDERSEN *et al.*, 1974].

This approach rests on an assumption that has a long history. In the classic tale of Gulliver's Travels in 1729, Jonathan Swift had shipwrecked Gulliver cast up on a strange shore and made captive of Lilliputians, who were little people one-twelfth his size. In one of the articles of the Agreement he signed when his freedom was granted, he was to receive a daily allotment of food and drink equivalent to that of 1,728 Lilliputian portions. This figure was arrived at by the court mathematicians who concluded that since Gulliver's stature exceeded theirs in the proportion of twelve to one, and since he was of the same shape and composition, his volume and mass would be a function of his height increase raised to the third power, that is, (12^3) or 1,728.

When this same rationale is applied to maximal aerobic power to dissociate it from size, height cubed (h^3) is used as a divisor as theoretically proposed by ASMUSSEN and HEEBÖLL-NIELSEN [1955] in formula 1, or since weight numerically approximates mass in the metric system, body weight (w) is customarily used as a divisor as in formula 2

$$\dot{V}O_2 \text{ max rel} = \dot{V}O_2 \text{ max min}^{-1}\,h^{-3} \quad (1)$$

$$\dot{V}O_2 \text{ max rel} = \dot{V}O_2 \text{ max min}^{-1}\,w^{-1} \quad (2)$$

It should be recognized that while physiologists have traditionally inserted units of weight for mass as in formulae 2, 4 and 5, in a physical sense the terms are distinct and different in dimension.

From this purely geometrical point of view, as discussed by ASMUSSEN and HEEBÖLL-NIELSEN [1955], WHO study group [1969], and ÅSTRAND and RODAHL [1970] all linear measures of the body (i.e. lengths, breadths, girths, skinfold thicknesses) have the dimension L; all areas including body surface area and strength, which is a function of the area of a cross-section of muscle tissue, have the dimension L^2; and all weights and volumes the dimension L^3. This geometrical model is based on the same assumption made by the Lilliputians that shape and composition are reasonably constant. This concept has been further interpreted by VON DOBELN and ERIKSSON [1973] who maintain that the body should be considered a dynamic system whereby maximal aerobic power is a volume L^3 per unit of time. From Newton's second law, in physiological systems, time has the dimension L, therefore maximal aerobic power should theoretically be proportional to $L^3 \times L^{-1} = L^2$. Using this rationale a size dissociated value for $\dot{V}O_2$max would take a form such as in formula 3 or 4.

$$\dot{V}O_2 \text{ max rel} = \dot{V}O_2 \text{ max min}^{-1} \text{ h}^{-2} \quad (3)$$

where h is stature or height of the subject

$$\dot{V}O_2 \text{ max rel} = \dot{V}O_2 \text{ max min}^{-1} \text{ w}^{-2/3} \quad (4)$$

where w is body weight raised to the power two-thirds, which is analagous to height squared when shape and composition are constant. Recent studies by ERIKSSON and KOCH [1972], ANDERSEN *et al.* [1974] and others have used height squared to express a size-dissociated value for aerobic power in growing children.

Another advocacy has emerged based on experimental evidence that in many species of homeotherms, metabolic rate does not increase to the surface law of body weight to the power two-thirds but is more closely proportional to body weight to the power three-quarters. KLEIBER's observation in 1932 and his reiteration in 1961 and 1972 have recently been ascribed a theoretical explanation by MCMAHON [1973] who claims elastic criteria impose limits on biological proportions and metabolic rates. These views would have a size dissociated value for maximal aerobic power expressed as follows:

$$\dot{V}O_2 \text{ max rel} = \dot{V}O_2 \text{ max min}^{-1} \text{ w}^{-3/4} \quad (5)$$

where w is body weight. In the formula below $h^{-2.25}$ is the geometrical height identity for $w^{-3/4}$.

$$\dot{V}O_2 \text{ max rel} = \dot{V}O_2 \text{ max min}^{-1} \text{ h}^{-2.25} \quad (6)$$

Thus, to dissociate maximal aerobic power from size, we have essentially three different approaches with theoretical backing which deserve attention:

Ht^3 (as proposed by ASMUSSEN and HEEBÖLL-NIELSEN [1955] with a geometrical weight identity of Wt^1;

Ht^2 (as proposed by VON DOBELN and ERIKSSON [1973] with a geometrical weight identity of $Wt^{2/3}$;

$Ht^{2.25}$ with a geometrical weight identity equivalent to MCMAHON's [1973] and KLEIBER's [1961] $Wt^{3/4}$.

The preceding formulae were subject to test by longitudinal $\dot{V}O_2$max data obtained from the Saskatoon Child Growth and Development Study. This Study was a 10-year longitudinal investigation of growth which originated in 1964 and terminated in 1973. The major purpose of the Saskatchewan

Study was to evaluate changes that occurred in the performance of boys and girls from age 7 years to 16 years in a variety of structural and functional measures. Previous published reports have dealt with speed of reaction and movement – CARRON and BAILEY [1973], strength – CARRON and BAILEY [1974], and motor performance – ELLIS *et al.* [1975]. The present study presents data from longitudinal observations on maximal aerobic power of boys during prepubertal and circumpubertal years.

Method of Investigation

Subjects

As discussed in previous publications by CARRON and BAILEY [1973, 1974], the original sample in the Saskatchewan Study consisted of 207 7-year old boys who were randomly selected on a stratified socioeconomic basis from the elementary school system in the City of Saskatoon. Subjects in the study were not exposed to any special program of activity other than the yearly test. Over the 10-year duration of the study a number of subjects withdrew for a variety of reasons, some subjects were missing in certain years, and some subjects failed to meet established criteria for the exercise test in certain years. Only subjects for whom complete data was available over an 8-year period were used in the present analysis. This yielded a longitudinal sample of 51 boys who were studied annually over the age range 8–15 years with respect to changes in $\dot{V}O_2$ max from year to year. Clearly these subjects were probably above average in fitness due to the bias that is inherent in a longitudinal study where disinterested subjects tend to drop out. Data when the subjects were 7 years old was not used in the present study because of the first time difficulty in pushing youngsters to maximum on the treadmill test. As well, 16-year-old data was not used because some of the boys had reached mature height and showed no increase in linear dimensions between 15 and 16 years of age, thereby making their data inappropriate for the purposes of this investigation.

Exercise Test

The treadmill test involved the minute-to-minute collection of data from a subject for 5 min of standing preexercise, followed by continuous exercise to exhaustion on a motor-driven treadmill at 0% grade with stepwise increases in speed (3 min at 3 mi/h, 3 min at 6 mi/h, 3 min at 9 mi/h, and if necessary at 12 mi/h until exhaustion) and 10 min of standing recovery. While ÅSTRAND [1971] states that the treadmill is preferred in studying children under the age of 12, the protocol followed did have the limitation that subjects systematically exercised longer as they became older (table I). Heart rate was monitored by means of chest lead electrocardiogram (ECG) fluid column electrodes, minute ventilation ($\dot{V}_1$) using a high speed dry gas meter (Parkinson-Cowan CD4) on the inspired side, and oxygen uptake ($\dot{V}O_2$) by analysis of the mixed air (Beckman E2 O_2 analyzer and Godart Capnograph CO_2 analyzer). To minimize the effect of circuit resistance a breathing valve of the type described by BANNISTER and CORMAK [1954] but with a larger inner diameter was used (dead air space 120 cm^3). At a constant flow rate of 250 liters/min the maximum resistance of the circuit was 12 mm of H_2O.

Table I. Means and standard deviations of 51 boys studied longitudinally over 8 years [Saskatoon Child Growth and Development Study]

Age at test occasion	Height, cm	Weight, kg	Treadmill run, sec	$\dot{V}O_2$ max l/min^{-1}	Terminal heart rate
8.07	127.9	25.9	448	1.46	193.4
0.26	5.1	3.3	45	0.20	10.2
9.06	133.4	28.9	457	1.72	196.2
0.29	5.1	3.7	72	0.22	10.8
10.05	138.7	32.0	471	1.82	195.5
0.30	5.3	4.3	56	0.23	99.0
11.05	143.8	34.8	489	1.96	197.6
0.30	5.6	4.8	58	0.25	8.9
12.04	148.9	38.5	501	2.18	195.3
0.29	6.2	5.7	70	0.33	8.9
13.03	155.3	43.4	536	2.39	198.3
0.29	7.6	7.7	67	0.39	7.9
14.05	162.8	50.0	563	2.73	197.3
0.29	7.9	8.9	74	0.53	9.8
15.03	169.9	55.9	575	2.94	197.3
0.29	7.1	8.5	81	0.50	10.2

Data Analysis

Although the treadmill test protocol met the requirements for a test of maximal aerobic power as discussed by ÅSTRAND [1971], not all subjects ran to exhaustion due to the vagaries of childhood motivation. On the assumption that maximal values were not likely to be attained unless the subjects had run at least through 1 min at the 9 mi/h speed ($\geq$7 min of exercise), this value was used as the minimum criteria for defining maximal exertion. The mean terminal heart rate values, in excess of 193 beats/min for each age level, as shown in table I, indicate that the subjects had in fact exercised at levels very close to maximum.

In all, 51 subjects met the minimum requirement of at least 420 sec of treadmill running on each of eight successive measurement occasions and to this extent they were a select sample.

Results and Discussion

Means and standard deviations for the 51 subjects for age of measurement occasion, height, weight, terminal exercise heart rate, treadmill run time, and maximal oxygen uptake in liters per minute are summarized in table I.

As shown in the summary table II, considerable differences were reported in maximal oxygen uptake values in cross-sectional treadmill studies in boys

Table II. Comparative maximal aerobic power values for boys (treadmill data)

Author	Age, years	Subjects n	Height, cm	Weight, kg	VO_2 max, l/min	VO_2 max, ml/kg/min	Max HR
I. Cross-sectional studies							
Robinson	6.1	4		21.0	0.98	46.7	
[1938] USA	10.4	9		30.0	1.56	52.1	
	14.1	9		55.8	2.63	47.1	
	17.4	11		68.5	3.61	52.8	
Morse *et al.*	10–12					48.2	
[1949] USA	13					44.8	
	14–17					50.9	
Åstrand	4–6	10	113.5	20.8	1.01	49.1	203
[1952] Sweden	7–9	12	135.0	30.7	1.75	56.9	208
	10–11	13	145.4	36.5	2.05	56.1	211
	12–13	19	154.4	43.6	2.46	56.5	205
	14–15	10	171.8	59.5	3.53	59.5	203
	16–18	9	176.9	64.1	3.68	57.6	202
Cerretelli *et al.*	8–10				1.50	48.0	185
[1963] Italy	10–12				1.69	48.7	189
	12–14				2.27	49.2	191
	14–15				3.09	51.3	192
	16–18				3.29	52.2	185
Metz *et al.*	12–13	30		63.1	2.55	50.9	199
[1970] USA	14–15	30		66.5	3.09	53.3	196
Matsui *et al.*	12	10	149.9	40.0	1.96	49.1	209
[1971] Japan	13	19	152.7	44.4	2.02	45.7	209
	14	16	158.9	47.4	2.31	50.7	205
	15	20	163.9	53.3	2.65	49.7	205
	16	15	166.3	56.0	3.03	55.4	201
	17	16	168.0	60.0	3.12	52.0	204
	18	10	167.1	58.2	3.18	55.6	209
Ikai *et al.*	8–10	19	129.6	26.2	1.31	50.0	192
[1972] Japan	10–11	18	134.5	30.8	1.50	49.1	193
	11–12	21	140.5	34.6	1.77	51.3	197
	12–13	14	150.2	42.4	1.90	44.9	184
	13–14	21	154.3	45.4	2.13	47.5	191
	14–15	19	160.1	47.7	2.35	49.5	191
	15–16	13	164.0	51.6	2.47	48.1	198
II. Longitudinal Studies							
Šprynarová *et al.*	10.9	114	144.5	37.1	1.77	48.0	196
[1965] Czech	11.9	114	149.7	40.7	2.06	50.7	198
	12.9	114	155.1	44.8	2.25	50.4	197
Hermansen *et al.*	10.5	20		36.0	1.96	54.3	206
[1971] Norway	11.5	20		39.9	2.17	54.7	203
	12.5	20		43.6	2.52	58.1	203

Table III. Relative VO_2 max values calculated according to three proposals for size dissociation (determinations based on actual data from a longitudinal sample of 51 boys over an 8-year period)

Age	Height, cm	Weight, kg	$\dot{V}O_2$ max, l/min	$\propto L^3$		$\propto L^2$		$\propto L^{2.25}$	
				w	h^3	$w^{2/3}$	h^2	$w^{3/4}$	$h^{2.25}$
8	127.9	25.9	1.46	56.4	56.4	56.4	56.4	56.4	56.4
9	133.4	28.9	1.72	59.5	58.6	61.8	61.1	61.2	60.4
10	138.7	37.0	1.82	56.9	55.1	61.1	59.8	60.0	58.6
11	143.8	34.8	1.96	56.3	53.3	62.2	59.9	60.7	58.2
12	148.9	38.5	2.18	56.6	53.4	64.7	62.1	62.6	59.8
13	155.3	43.4	2.39	55.1	51.6	65.4	62.6	62.7	59.7
14	162.8	50.0	2.73	54.6	51.1	68.0	65.1	64.4	61.3
15	169.9	55.9	2.94	52.6	48.5	68.0	64.4	63.8	59.9

w = $\dot{V}O_2$ max ml min^{-1} kg^{-1}. All other values have been scaled in arbitrary units for comparative purposes starting from a similar 8-year-old reference point and using the scaling factors listed below:

h^3 8.08235 E+07

h^2 6.31928 E+05 $w^{2/3}$ 3.38163 E+02

$h^{2.25}$ 2.12513 E+06 $w^{3/4}$ 4.43508 E+02

spanning four or more successive age levels. As expected, all of these studies showed increased oxygen uptake in liters per minute with increasing age. However, in terms of liters per minute per kilogram of body weight, the traditional way of reporting aerobic power, the findings were equivocal and no consistent pattern with increasing age emerged.

The majority of longitudinal studies on maximal aerobic power are experimental in nature involving training stimuli of some sort. While these studies provide important information on aerobic capacity, the effects attributable to growth and training are often inextricable. Table II summarizes treadmill data by ŠPRYNAROVÁ and PAŘÍZKOVÁ [1965] and HERMANSEN and OSEID [1971] on 'normal' boys measured annually on three occasions. As in the Saskatchewan data, gross uptake increased with age. In terms of milliliters per kilogram of body weight per minute, ŠPRYNAROVÁ and PAŘÍZKOVÁ [1965] report relatively stable values over the age range 10.9, 11.9 and 12.9 in Czech children with a slight increase in the last 2 years. HERMANSEN and OSEID's [1971] study of Norwegian children at ages 10.5, 11.5 and 12.5 report stable

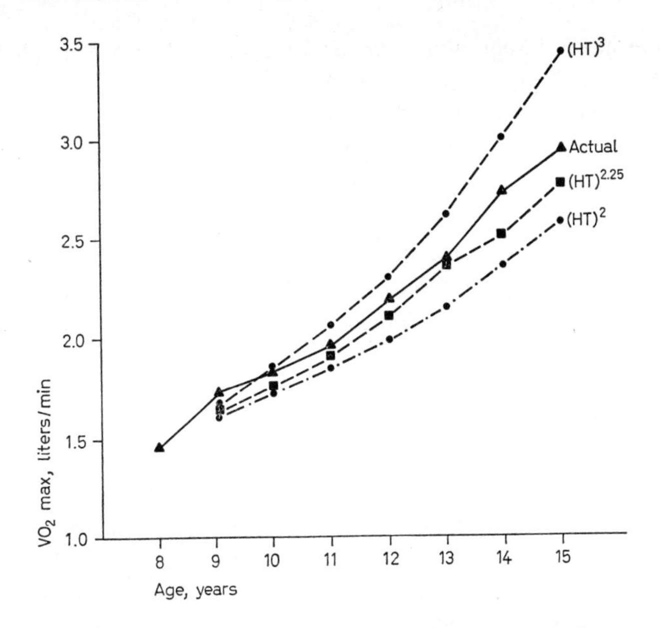

Fig. 1. Obtained mean curve for maximal aerobic power ($\dot{V}O_2$ max liters/min) from age 8 to 15 years in 51 boys studied longitudinally compared with postulated curves based upon the mean increase in the linear dimensions of the boys raised to the power 3, 2 and 2.25.

values over the first 2 years with a large increase in the final year. In the Saskatchewan data over a larger age range, with the exception of an elevated value at age 9, values for $\dot{V}O_2$max in ml/kg/min were reasonably stable from 8 through 12 years with a successive decline thereafter (table III). In magnitude the Saskatchewan values like the Norwegian values are relatively high compared to the results of some cross-sectional studies. This is probably a reflection of the built-in selectivity inherent in longitudinal as opposed to cross-sectional studies where bias may be introduced into the sample with the attrition of disinterested subjects over extended time periods.

Figure 1 shows the absolute increase in $\dot{V}O_2$max in liters/min for the longitudinal sample of 51 boys over the ages of 8–15 years. This distance curve based on actual values may be compared to three theoretical curves which have been drawn on the basis of postulated increases in $\dot{V}O_2$max according to height raised to the exponents 3, 2 and 2.25. These are the height identities for weight to the 1, 2/3 and 3/4 exponents as per ASMUSSEN and HEEBÖLL-NIELSEN [1955], VON DOBELN and ERIKSSON [1973], and KLEIBER [1961] respectively.

The actual curve most closely approximates the curve of KLEIBER in which postulated increases in $\dot{V}O_2$max in liters/min were based on proportional increases in the linear dimensions to the 2.25 exponent. The exponent which best describes the actual curve is 2.46 which is considered to be quite close to KLEIBER's observations from the animal kingdom.

In order to explore the various proposals for a size-dissociated value for maximal aerobic power over time, the theoretical formulae suggested by various investigators and outlined previously, were solved according to the size values obtained from 51 boys studied longitudinally over 8 years. The means of these $\dot{V}O_2$max rel values, or products of the means and a scale factor are summarized in table III.

If shape, proportion and composition were constant with age and we make the assumption that there was no systematic change in the exercise environment of the growing youngsters, the best relative maximal aerobic power value would yield a horizontal graphline with age. From table III it can be seen that by using w or its identity h^3 there is a descent in $\dot{V}O_2$max at the older ages. Using these divisors the implication would be that relative maximal aerobic power and fitness declined at the older ages. The divisors $w^{2/3}$ and its geometrical height analogue of h^2 yielded an increasing relative maximal aerobic power value with age. The implication from these divisors is that relative fitness increases with age. The remaining conventional size adjustments of $h^{2.25}$, which is the geometrical identity for $w^{3/4}$, also yielded increasing values with age, although to a lesser degree. While none of the conventional size adjustments fully satisfied the criterion for obtaining a perfectly horizontal baseline with age, it is clear that the most stable size dissociated value for maximal aerobic power over time resulted from using the divisor $h^{2.25}$. Thus, the height identity for KLEIBER's three-fourth power of body weight rule would seem to be the best theoretical representation of metabolic body size for growing boys in this study.

At least three reasons might be advanced to account for the fact that the obtained data did not exactly fit KLEIBER's rationale. (1) The assumption of similar shape and composition over the entire age range studied may not be tenable. (2) The assumption of constancy in the exercise environment of the sample over age is questionable and virtually impossible to control. (3) The assumption that maximal oxygen uptake is one of a family of metabolic parameters which are dimensionally related to size, may not be tenable. As pointed out in personal communication with ULTSCH [1973] who recently reviewed the interspecies literature on the topic of body size and oxygen exchange capacity… 'there may be a different relationship between maximal metabolic rate and body weight than between basal metabolic rate and body weight. But the data necessary for a general statement is lacking'.

Hopefully, data from the present longitudinal study will serve to clarify some of the concepts concerning body size and maximal oxygen exchange capacity in children. It is doubtful if answers will come from cross-sectional studies because as DEHN and BRUCE [1972] point out, cross-sectional studies may misrepresent trends in $\dot{V}O_2$max during growth. Also, if different children are measured at different ages (as in a cross-sectional study) the assumptions inherent to the problem become even harder to accept.

At this point, data from the present study is interpreted as broadly substantiating the dimensional concepts of KLEIBER. The use of height to the power 2.46 derived from the Saskatchewan data more closely approximates $h^{2.25}$, which is the height analogue for KLEIBER's metabolic rule of $w^{3/4}$, than it does other proposals. The use of $\dot{V}O_2$max rel = $\dot{V}O_2$max min^{-1} $h^{-2.25}$ as a size-dissociated value for maximal aerobic power in serial observations in growing children can therefore be supported on empirical grounds as well as on theoretical grounds as proposed by KLEIBER [1961, 1972]. The use of $h^{2.25}$ rather than the weight identity of $w^{3}/^{4}$ is conceptually preferable to the physical scientist, and also has the advantage of being largely independent of extrinsic environmental factors. The findings of this study should have some practical significance in studying the influence of training on children where the growth component must be factored out. Moreover, until more definitive information is available, these findings perhaps serve as the best available evidence of the dimensional relationship of size with maximal aerobic power on a fairly large sample of boys studied longitudinally during the growth years.

Summary

The yearly administration of a treadmill test to 51 boys from the age of 8 to 15 years yielded longitudinal $\dot{V}O_2$ max data which was analyzed to evaluate different theoretical approaches which have been proposed to dissociate maximal aerobic power from size. Increases in linear dimensions and weight during growth were compared to increases in $\dot{V}O_2$ max in liters/min to determine the dimensional relationship between body size and oxygen exchange capacity. Results of the study are broadly interpreted as substantiating Kleiber's thesis that the three-fourth power of body weight, with a geometrical height identity of $h^{2.25}$, is representative of metabolic body size. The use of the formula $\dot{V}O_2$ max rel = $\dot{V}O_2$ max min^{-1} $h^{-2.25}$ is seen as being worthwhile in studying the influence of training and other extrinsic environmental conditions on children where the growth component must be factored out.

References

ANDERSEN, K. L.; SELIGER, V.; RUTENFRANZ, J., and MOCELLIN, R.: Physical performance capacity of children in Norway. Eur. J. appl. Physiol. *33:* 177–195 (1974).

ASMUSSEN, E. and HEEBÖLL-NIELSEN, K.: Physical performance and growth in children. Influence of sex, age, and intelligence. J. appl. Physiol. *8:* 371–380 (1955).

ÅSTRAND, P. O.: Experimental studies of physical working capacity in relation to sex and age, pp. 103–123 (Munksgaard, Copenhagen 1952).

ÅSTRAND, P. O. and RODAHL, K.: Textbook of work physiology, pp. 321–340 (McGraw-Hill, New York 1970).

ÅSTRAND, P. O.: Definitions, testing procedures, accuracy, and reproducibility. Acta. paediat., Stockh. *271:* suppl., pp. 9–12 (1971).

BANNISTER, R. G. and CORMAK, R. S.: Two low resistance low dead space respiratory values. J. Physiol., Lond. *124:* 4 (1954).

CARRON, A. V. and BAILEY, D. A.: A longitudinal examination of speed of reaction and speed of movement in young boys 7 to 13 years. Hum. Biol. *45:* 663–681 (1973).

CARRON, A. V. and BAILEY, D. A.: Strength development of boys from 10 through 16 years. Monogr. Soc. Res. Child Dev., Serial No. 157, No. 4 (1974).

CERRETELLI, P.; AGHEMO, P., and REVILLE, E.: Morphological and physiological observations in school children in Milan. Med. Sport *3:* 731–748 (1963).

DEHN, M. M. and BRUCE, R. A.: Longitudinal variations in maximal oxygen intake with age and activity. J. appl. Physiol. *33:* 805–807 (1972).

ELLIS, J. D.; CARRON, A. V., and BAILEY, D. A.: Physical performance in boys from 10 through 16 years. Hum. Biol. *47:* 263–281 (1975).

ERIKSSON, B. O. and KOCH, G.: Effect of physical training on hemodynamic response during submaximal and maximal exercise in 11–13 year old boys. Acta physiol. scand. (1972).

HERMANSEN, I. and OSEID, S.: Direct and indirect estimation of maximal oxygen uptake in pre-pubertal boys. Acta paediat., Stockh. *217:* suppl., pp. 18–23 (1971).

IKAI, M. and KITAGAWA, K.: Maximum oxygen uptake of Japanese related to sex and age. Med. Sci. Sports *4:* 127–131 (1972).

KLEIBER, M.: Hilgardia *6:* 315 (1932).

KLEIBER, M.: The fire of life, pp. 177–216 (Wiley, New York 1961).

KLEIBER, M.: Body size conductance for animal heat flow and Newton's law of cooling. J. theor. Biol. *37:* 139–150 (1972).

MATSUI, H.; MIYASHITA, M.; MIURA, M.; AMANO, K.; MIZUTANI, S.; HOSHIKAWA, T.; TOYOSHIMA, S., and KAMEI, S.: Aerobic work capacity of Japanese adolescents. J. Sports Med. phys. Fitness *11:* 28–35 (1971)

MCMAHON, T.: Size and shape in biology. Science *199:* 1201–1204 (1973).

METZ, K. F. and ALEXANDER, J. F.: Estimation of maximal oxygen intake from submaximal work parameters. Res. Q. *42:* 187–193 (1971).

MORSE, M.; SCHULTZ, F. W., and CASSELS, D. E.: Relation of age to physiological responses of the older boy (10–17 years) to exercise. J. appl. Physiol. *1:* 683–709 (1949).

ROBINSON, S.: Experimental studies of physical fitness in relation to age. Arbeitsphysiologie *10:* 251–323 (1938).

ŠPRYNAROVÁ, Š. and PAŘÍZKOVÁ, J.: Changes in the aerobic capacity and body composition in obese boys after reduction. J. appl. Physiol. *29:* 934–937 (1965).

TENNEY, S. M.: Some aspects of the comparative physiology of muscular exercise in mammals. Circulation Res. *20:* suppl. 1, pp. 7–14 (1967).

ULTSCH, G. R.: A theoretical and experimental investigation between metabolic rate, body size and oxygen exchange capacity. Resp. Physiol. *18:* 143–160 (1973).

VON DOBELN, W. and ERIKSSON, B. O.: Physical training, growth and maximal oxygen uptake. Proceedings IV International Symposium on Paediatric Work Physiology, pp. 93–108 (Bar-Or, Wingate 1973).

WHO Scientific Groups: Optimal performance capacity in adults. Wld Hlth Org. Tech. Rep. Ser. *436:* 10–21 (1969).

Prof. Dr. DONALD A. BAILEY, University of Saskatchewan, College of Physical Education, *Saskatoon, S7N OWO* (Canada)

Medicine Sport, vol. 11, pp. 152–158 (Karger, Basel 1978)

Muscle Strength Development in the Pre- and Post-Pubescent Age

J. VRIJENS

Department of Physical Education, Rijksuniversiteit, Gent

During the last years, an increasing number of studies concerning the influence of intensive training for children has been published [GANDELMANN, 1964; GRIMM and KELLERMANN, 1966; WEIDEMANN, 1968; SZABO, 1969; SZÖGY and ZAMFIRESCU, 1971; ULBRICHT, 1971; RIECKERT and GABLER, 1972]. Most of the experiments focused their attention on cardio-vascular fitness. As to the effect of strength training in children, the findings of all authors are in general concordant. It seems that strength development is closely related to sexual maturation [HETTINGER, 1958, 1964]. Therefore, specific strength training can only be effective in the post-pubescent age [IWANOW, 1964, 1965; HEBBELINCK, 1964; BÖTTGER, 1964; BERGER, 1965; SZABO, 1969]. However, some authors reported that even during the childhood period, muscle strength can be developed [NOACK, 1956; AKKERMAN, 1963; GRIMM and RAEDE, 1967; ROHMERT, 1968]. This study therefore compared the effects of strength training in young children and in adolescent boys.

Methods

A group of children in the pre-pubescent age (A) and adolescents in the post-pubescent growing age (B) were submitted to a strength training program (table I).

Training Program

The subjects of group A were 16 boys with a mean age of 10 years, 5 months. The adolescent group (B) were 12 boys with a mean age of 16 years, 8 months.

Both groups were trained for strength by means of isotonic exercises. The following muscle groups were trained: arm flexors, arm extensors, leg extensors, leg flexors, abdominal and back muscles. The boys performed a circuit of 8 exercises once during each training session. For each exercise, the number of repetitions varied between 8 and 12,

Table I. Training schedule

(a) Trained muscle groups:	arm flexors	back muscles
	arm extensors	abdominal muscles
	leg extensors	
	leg flexors	
(b) Type of exercises:	concentric isotonic exercises	
(c) Intensity of load:	±75% of maximum load	
(d) Number of repetitions:	between 8 and 12	
(e) Number of training sessions:	3 times a week	
(f) Duration of training period:	8 weeks	

which corresponds a load of ±75% of IRM [VRIJENS, 1970]. The boys exercised three times a week for a total period of 8 weeks. During the first week, special care was paid to the correct execution of all exercises. Performances of each boy were recorded on score cards which allowed an individual control during the whole training period.

Test Methods

All boys were measured before and after the training period. The following parameters were used to evaluate the changes in muscular development and strength.

Anthropometric tests consisted of measurement of body weight, standing height, and arm and thigh circumferences. Skinfold thicknesses on the right side of the body were measured according to the method of ALLEN *et al.* [1956] over 10 different sites of the body. Mean thicknesses allowed indirect indication of adiposity.

Muscular development was studied by soft-tissue roentgenography. Measurements of cross-sectional surface area of muscle and fat at well-defined points were taken on the thigh and the upper arm.

Strength was tested with the HETTINGER [1958] apparatus. In a standard position, isometric (static) strength of six muscle groups was measured. Each test consisted of three maximal contractions and the best score was recorded. The following muscle groups were successively tested: arm flexors and extensors, leg extensors and flexors, abdominal and back muscles.

Results

The changes of static strength produced by the training program are listed in table II.

For the pre-pubescent group, no consistent pattern of strength change was noted. No improvements were obtained in the lower and upper extremities, whereas isometric strength of the trunk muscles increased significantly. In the group of adolescent boys, on the contrary, strength increased in all tested muscle groups. The variations in training effect in the pre-pubescent

Table II. Isometric strength scores before and after 8 weeks of training

	Pre-pubescent group (n:16)			Post-pubescent group (n:12)		
	$\overline{X}_1$	$\overline{X}_2$	diff.	$\overline{X}_1$	$\overline{X}_2$	diff
Arm flexors, kg	12 5	12 1	− 0.4	23.5	27.5	+ 4.0[b]
Arm extensors, kg	10.2	10.8	+ 0.6	15.4	20.4	+ 5.0[c]
Leg extensors, kg	21.1	20.1	− 1.0	47.2	56.1	+ 8.9[b]
Leg flexors, kg	14.5	14.8	+ 0.3	22.5	29.6	+ 7.1[a]
Abdomen, kg	16.9	23.0	+ 6.1[a]	43.1	54.1	+11.0[b]
Back muscles, kg	49.4	66.8	+17.4[a]	118.0	135.9	+17.9[b]

[a] Significant at 0.01 level; [b] significant at 0.02 level; [c] significant at 0.05 level.

Table III. Cross-sectional surface area of muscle and fat before and after 8 weeks of training

	Pre-pubescent group (n:16)			Post-pubescent group (n:12)		
	$\overline{X}_1$	$\overline{X}_2$	diff.	$\overline{X}_1$	$\overline{X}_2$	diff.
Thigh						
Muscle, cm²	101.8	103.0	+1.2	185.7	194.2	+8.5[b]
Fat, cm²	34.8	34.6	−0.2	42.1	37.2	−5.0[b]
Arm						
Muscle, cm²	25.4	26.0	+0.6	43.5	49.7	+6.2[a]
Fat, cm²	13.3	12.9	−0.4	13.2	12.4	−0.8

[a] Significant at 0.02 level; [b] significant at 0.05 level.

group was a rather surprising fact. The lack of strength development in the extremities of youngsters was also confirmed by roentgenographic measurements (table III).

The mean cross-sectional area of muscles in the arm and thigh remained unchanged. In the adolescent group, a significant increase of muscle mass in agreement to the isometric strength improvement could be observed.

The results of the anthropometric measurements of the two groups before and after training are listed in table IV.

Table IV. Anthropometric data before and after 8 weeks of training

	Pre-pubescent age (n:16)			Post-pubescent age (n:12)		
	$\bar{X}_1$	$\bar{X}_2$	diff.	$\bar{X}_1$	$\bar{X}_2$	diff.
Weight, kg	31.9	32.5	+0.6[a]	60.0	60.8	+0.8[b]
Height, cm	138.7	140.0	1.3[a]	173.2	173.5	+0.3
Mean skinfold thickness, mm	8.2	8.0	−0.2	8.6	8.2	−0.4
Arm girth, cm	19.2	19.5	+0.3[c]	25.5	26.5	+1.0[a]
Thigh girth, cm	41.2	41.2	–	50.5	51.5	+1.0[b]

[a] Significant at 0.01 level; [b] significant at 0.02 level; [c] significant at 0.05 level.

For the pre-pubescent group, mean body weight and height increased significantly. The increase in weight can be explained by growth.

For the adolescent boys, body weight and the thigh and arm circumferences were significantly changed. The increased weight and muscular cross-sectional area together with unchanged mean skinfold thicknesses and cross-sectional area of fat are indicators of muscular development.

Discussion

The response to strength training during growth is subject to distinct variations. Different authors have indicated the close relationship between physical maturation and strength development. Children have relatively less muscle mass than adults. In the pre-pubescent age, muscle weight is about 27% of the total weight. After sexual maturation, muscular development is influenced by androgenic hormones and the percentage of muscle weight is increased to ±40%.

Most authors therefore concluded that strength development can be most effective after sexual maturation [Johnson, 1961; Böttger, 1964; Iwanow, 1964; Hebbelinck, 1964; Berger, 1965; Matthias and Rüfung, 1966; Szabo, 1969; Hollman and Bouchard, 1970; Beunen *et al.*, 1972].

Some authors, however, reported that children can be trained for strength. Grimm and Raede [1967], in an experiment based upon dynamic

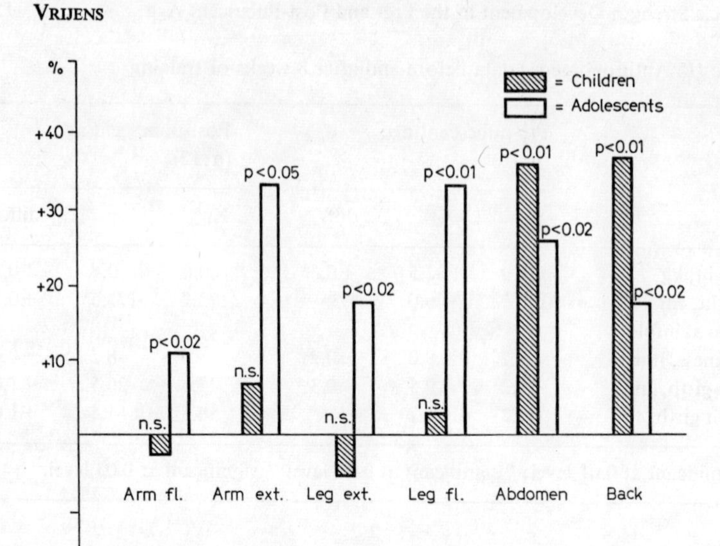

Fig. 1. Percentage of changes in strength after 8 weeks of training.

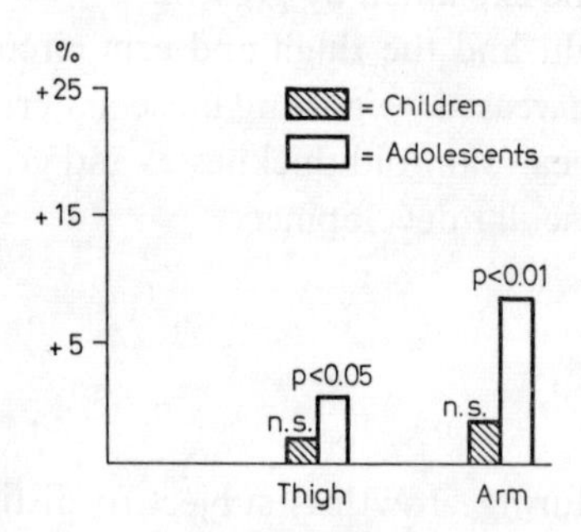

Fig. 2. Percentage of changes in muscle surface area after 8 weeks of training.

muscle strength, and ROHMERT [1968], with an isometric strength program, attained significant changes in muscle strength in pre-pubescent subjects.

Our studies, however, point to variations in training response during growth. In childhood (fig. 1), the absolute strength of the extremities is not changed by specific exercises.

The absolute trunk muscle strength of pre-pubescent boys surprisingly increased significantly by strength training. The difference in training response in childhood and the late adolescent age is not only distinct for strength scores but is also revealed by roentgenographic measurements (fig. 2).

The positive effect of strength training in abdominal and back muscle of children can be considered as an important contribution to the prevention of deficiencies in body posture during growth [MATTHIAS and RÜFUNG, 1966].

References

AKKERMAN, S.: Das Training des kindlichen und jugendlichen Turners, Med. Sport *III:* 112–114 (1963).

ALLEN, T.H.; FENG, M.T.; CHEN, K.B.; HUANG, T.F.; CHANG, C., and FANG, H.S.: Prediction of total adiposity from skinfolds and the curvilinear relationship between external and internal adiposity. Metabolism *3:* 346 (1956).

BERGER, J.: Zu einigen Fragen des Muskelkraft-Trainings im Kindes- und Jugendalter. Theor. Prax. Körperkult. *14:* 1083–1092 (1965).

BEUNEN, G.; OSTYN, M.; RENSON, R.; SIMONS, J.; SNAKERS, F.; GERVEN, D. VAN, and WILLEMS, G.J.: Skeletleeftijd en fysische ontwikkeling bij twaalfjarige jongens. Belg. Arch. Soc. Geneesk. *30:* 102–119 (1972).

BÖTTGER, W.: Fragen der Belastungswirkungen sportlichen Trainings auf den im Wachstum befindlichen Organismus. Med. Sport *IV:* 185–187 (1964).

CLARKE, H.H. and DEGUTIS, E.W.: Relationships between standing broad jump and various maturational anthropometric and strength tests of 12-year-old boys. Res. Q. *35:* 258–264 (1964).

GANDELMANN, A.B.: Über die physiologische Grundlage der Normierung von Trainingsbelastungen bei Kindern im Schulalter. Theor. Prax. Körperkult. *2:* 114–119 (1964).

GRIMM, D. und KELLERMANN, S.: Durch Kreisbetrieb zu einer besseren allgemeinen Leistungsfähigkeit bei 8- und 9jährigen Schülern. Med. Sport *6:* 170 (1966).

GRIMM, D. und RAEDE, H.: Erfolgreiche Anwendung des Kreisbetriebs in einer 3. Klasse. Theor. Prax. Körperkult. *16:* 333–342 (1967).

HEBBELINCK, M.: De invloed van lichamelijke opvoeding en de sport op jongeren van 10 tot 14 jaar. Sport *7:* 232–236 (1964).

HETTINGER, T.: Die Trainierbarkeit menschlicher Muskeln in Abhängigkeit von Alter und Geschlecht. Int. Z. angew. Physiol. *17:* 371–377 (1958).

HETTINGER, T.: Isometrisches Muskeltraining (Thieme, Stuttgart 1964).

HOLLMANN, W. und BOUCHARD, C.: Untersuchungen über die Beziehungen zwischen chronologischem und biologischem Alter zu spiroergometrischen Messgrössen, Herzvolumen, anthropometrischen Daten und Skelett. Z. Kreislaufforsch. *59:* 160–176 (1970).

IWANOW, S.M.: Medizinische Probleme des Kinder- und Jugendsports. Theor. Prax. Körperkult. *3:* 1106–1114 (1964).

IWANOW, S.M.: Belastungsnormen und sportärztliche pädagogische Kontrolle im Kinder- und Jugendsport. Theor. Prax. Körperkult. *14:* 152–159 (1965).

JOHNSON, A.: Influence of pubertal development on response to exercise. Res. Q. *27:* 182 (1961).

KEUL, H.; REINDELL, H. und ROSKAMM, H.: Trainingsauswirkungen beim Jugendlichen. Sportarzt *12:* 254–258 (1961).

Matthias, H.H. und Rüfung, A.: Wachstum und Wachstumsstörungen des Haltungs- und Bewegungsapparates im Jugendalter (Karger, Basel 1966).

Noack, H.: Theor. Prax. Körperkult. *5:* 885 (1956).

Peters, H.: Die physische Leistungsfähigkeit und Belastbarkeit der Unterstufenschüler. Theor. Prax. Körperkult. *17:* 248–257 (1968).

Rieckert, H. und Gabler, H.: Der Trainingseffekt einer täglichen Sportstunde auf das körperliche Leistungsvermögen von 11–12jährigen Schülern. Sportarzt Sportmed. *23:* 21–25 (1972).

Rohmert, W.: Rechts-Links-Vergleich bei sometrischem Armmuskeltraining mit verschiedenem Trainingsreiz bei achtjährigen Kindern. Int. Z. angew. Physiol. *26:* 363–393 (1968).

Szabo, S.: Die Zunahme von Herzvolumen und Leistungsfähigkeit als Folge eines regelmässigen Trainings beim Schwimmen im Pubertätsalter. Med. Sport *IX:* 280–282 (1969).

Szögy, A. und Zamfirescu, N.R.: Zur Beziehung zwischen spiroergometrischen Messgrössen und dem Herzvolumen bei jugendlichen Sportlern unter besonderer Berücksichtigung des Alters und des Körperwachstums. Int. Z. angew. Physiol. *29:* 328–336 (1971).

Tanner, J.M. and Whitehouse, R.M.: The Harpenden skinfold caliper. Am. J. Physiol. *13:* 743 (1955).

Ulbrich, J.: Individual variants of physical fitness in boys from the age of 11 up to maturity and their selection for sport activities. Med. Sport *24:* 118–135 (1971).

Vrijens, J.: Vergelijkende studie van dynamische en statische contractie in verband met de spierkracht. Tijdschr. Lichamelijke Opvoeding *35:* 31–38 (1970).

Weidemann, H.: Herz und Kreislauf bei jugendlichen Sportlern. Lichamelijke Opvoeding *56:* 3–6 (1968).

Dr. J. Vrijens, Department of Physical Education, Rijksuniversiteit, *Gent* (Belgium)

Medicine Sport, vol. 11, pp. 159–166 (Karger, Basel 1978)

Investigation into the Effects of Two Extra Physical Education Lessons per Week during one School Year upon the Physical Development of 12- and 13-Year-Old Boys[1]

HAN C. G. KEMPER, ROBBERT VERSCHUUR, KOOS G. A. RAS, JAN SNEL, PAUL G. SPLINTER and LOUIS W. C. TAVECCHIO

University of Amsterdam, Jan Swammerdam Institute, Amsterdam

Introduction

The reason why this investigation has been set up was the wish, repeatedly expressed by Dutch education authorities, to increase the number of weekly lessons in physical education (p-e) in schools from 3 to 5. In the Proposed Curriculum for the Government Schools [1968], one finds the aims of p-e expressed as follows: (1) to produce a favorable influence on development of the body; (2) to promote good bearing and stature; (3) to increase the willingness and ability to participate in physical activity; (4) to stimulate teamwork; (5) to form good health habits; (6) to become acquainted with valuable forms of active recreation for one's leisure time.

In a pilot study, the influence was studied of 5 p-e lessons per week compared with 3 lessons per week in the course of a whole school year upon the physiological and psychological characteristics of 12- and 13-year-old schoolboys [KEMPER, 1973]. The results did not suggest that there was any clear improvement in physical development. Physical skill levels, however, became increasingly higher in the experimental class.

In order to retest the results of this pilot study, it was decided to replicate the former experiment.

[1] This is a combined project of the Laboratory of Psychophysiology (Prof. Dr. P. VISSER) and the Coronel Laboratory (Prof. Dr. R. L. ZIELHUIS), supported by a grant of the Foundation for Educational Research (SVO) and the Department of Health and Environmental Hygiene in The Hague (Project No. 0185).

Methods

Subjects were boys of the 4 first forms of a secondary school in Amsterdam, St. Ignatius College. From 82 boys, 12 dropped out. At the beginning of the school year 1971/1972, their mean chronological age was 12.5 ($\pm$0.4) years. Each class had its own teacher of p-e, who was also the mentor of that class. Two classes were assigned by lot as experimental and two classes as control group.

The independent variable was the frequency of lessons per week in p-e. Duration, intensity and content were held constant. The usual number of 3 lessons of p-e a week were given to the control group and 5 lessons of p-e a week to the experimental group. The 2 extra lessons of physical education had been added to the timetable in such a way that the experimental group received a total of 34 instead of 32 lessons week.

At the beginning (pretest) and at the end of the school year (posttest), the following groups of dependent variables were measured in a pretest-posttest control group design.

1. Anthropometric Variables, Measuring Body Build and Body Composition

Height, weight, and circumference, breadth and skinfold measurements were taken to evaluate the influence on skeletal, fat and muscle mass. The weight-for-height relationship in these subjects was compared with Dutch boys [VAN WIERINGEN *et al.*, 1968]. Values lay within the normal range, between the 10th and 90th percentile (fig. 1).

2. Physiological Variables

Aerobic power, measured as physical working capacity (W_{170}) [WAHLUND, 1948], was determined on a bicycle ergometer (fig. 2).

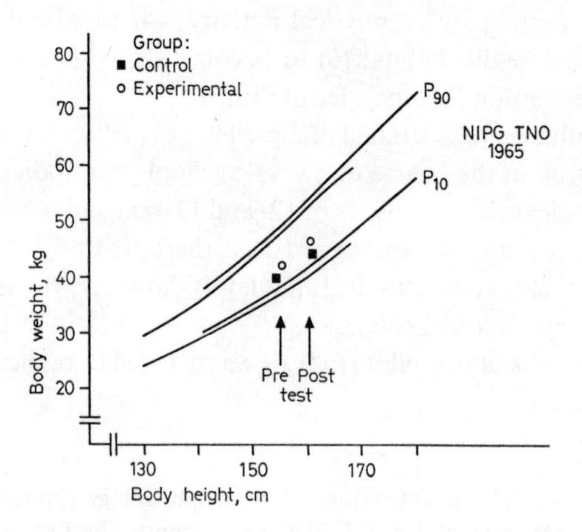

Fig. 1. Comparison of weight-for-height relation in Dutch boys (1965) with our subjects.

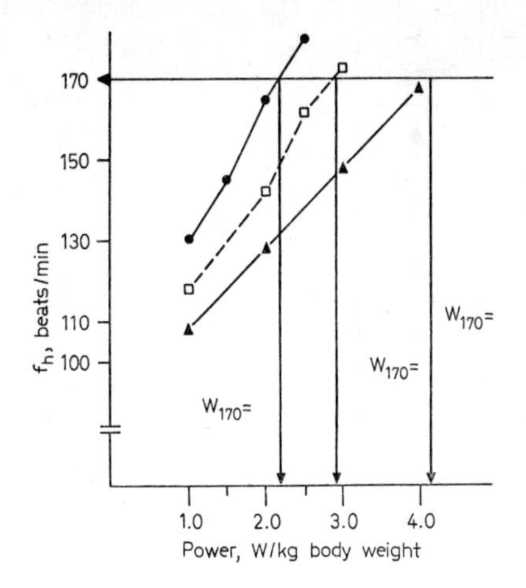

Fig. 2. Determination of W_{170} on bicycle ergometer test in 3 subjects.

The exercise test consisted of 4 load periods of 3 min each. The load was related to body weight. All subjects started with a load of 1.0 W/kg body weight. Each successive period the load was increased. The increase in load was determined on the basis of the heart rate reached in the last minute of the preceding load period (fig. 3).

For the evaluation of other aspects of the physical performance, simple performance tests were employed. The choice of the tests was based on the results of the only study that applied factor analysis to test scores of boys of the same age [SIMONS *et al.*, 1970]. Measures included: explosive arm strength (handgrip), explosive leg strength (vertical jump), muscular endurance (bent arm hang), flexibility (sit and reach), speed of arm movement (plate tapping), running speed (50-meter shuttle run) and aerobic power (12-min run-walk test).

The ventilation performance was measured as FEV%.

Period	Time	Intensity							
1	3 min	1.0 W/kg							
		Choice of following loads depends on mean heart rate (f_h) in the last minute of the preceding load:							
		$f_h \geq 120$				$f_h < 120$			
2	3 min	1.5 W/kg				2.0 W/kg			
		$f_h \geq 140$		$f_h < 140$		$f_h \geq 140$		$f_h < 140$	
3	3 min	2.0 W/kg		2.5 W/kg		2.5 W/kg		3.0 W/kg	
		$f_h \geq 160$	$f_h < 160$	$f_h \geq 160$	$f_h < 160$	$f_h \geq 160$	$f_h < 160$	$f_h \geq 160$	$f_h < 160$
4	3 min	2.5 W/kg	3.0 W/kg	3.0 W/kg	3.5 W/kg	3.0 W/kg	3.5 W/kg	3.5 W/kg	4.0 W/kg

Fig. 3. Procedure of the loading pattern on the bicycle ergometer.

No.	Factor	Representative variable	Direction
1	Ventilation	FEV %	↑
2	Muscle mass	corr. upper arm diam.	↑
3	Fat mass	fat %	↓
4	Speed of limbs	plate tapping	↑
5	Running speed	50-meter shuttle run	↑
6	Muscular endurance	bent arm hang	↑
7	Explosive arm strength	hand grip	↑
8	Aerobic power	W_{170}	↑
9	Perform. in phys. educ.	score skill tests	↑

Fig. 4. Definitive hypotheses. Arrows indicate in which direction the effect will be expected.

3. Skill Variables, Measuring Progress in Performance of Physical Education in a Narrower Sense

Objective performance tests were developed to measure progress in skills. Every test was scored on a 5-point scale. A total of 31 tests was used. To obtain an impression of the progress, the school year was divided into 4 periods. The sum of the scores on the skill tests was considered as the school grade in p-e for that period (fig. 7).

Based upon a review of the literature [KEMPER *et al.*, 1974], 9 hypotheses were formulated about the effect of the independent variable upon the dependent variables (fig. 4). The arrows indicate in which direction the effect was expected.

Although an increase from 3 to 5 lessons a week seemed sufficient to find effects of extra lessons of p-e, there may have been interfering variables that could mask these possible effects.

The following were considered to be potential interfering variables: (1) biological age, measured on pretest as skeletal age by X-ray photography of hand and wrist; (2) habitual physical activity, measured by pedometers and a questionnaire.

On pretest, the range of skeletal age, averaging 12.8 years ($\pm$0.8) was 3.1 years. On pretest, the range of chronological age, averaging 12.5 years ($\pm$0.4), was 1.8 year. Thus, the range of the skeletal age of these pupils was two times the range of chronological age.

With pedometers attached to the waist of the subjects, information about the amount of physical activity of the boys during leisure time was gathered. From the pedometer scores, remarkable interindividual differences in habitual physical activity could be demonstrated. In some periods, these differences were 6, 7 or even 10-fold.

α = Independent variable
γ = Skeletal age
β = Habitual physical activity

Test 1. H_0: $\alpha = \beta^{(e)} = \beta^{(c)} = \gamma = 0$
Test 2. H_0: $\alpha = 0$ and $\beta^{(e)} = \beta^{(c)}$
Test 3. H_0: $\beta^{(e)} = \beta^{(c)}$
Test 4. H_0: $\beta^{(e)} = 0$ and $\beta^{(c)} = 0$

Fig. 5. Four tests used in the analysis of covariance. H_0 = Null hypothesis. (e) = experimental and (c) = control group.

From the questionnaire, it was concluded that 30.8% of the total time was spent for transportation, 16.1% for organized activities and 53.1% for all other activities during leisure time.

The systematic differences between experimental and control group were analyzed by the following statistical steps: (a) The dependent variables were divided into groups that were supposed to be more or less independent from each other. (b) The grouped pretest data were factor-analyzed for both experimental and control group. Further, the representative variables found in both groups were used to formulate the hypotheses (fig. 4). (c) For each of the hypotheses, the mean difference scores of experimental and control group were compared by way of analysis of covariance, while making allowance for the influence of the independent variable (α), the habitual physical activity (β), the skeletal age (γ), and interaction of α and β (fig. 5).

Four tests were applied. If test 1 is significant, there is a difference between experimental and control group caused either by the extra lessons (α), habitual physical activity (β), or biological age (γ). If test 3 is not significant, it can be said that the difference between experimental and control group is caused only by the additional physical education. Test 4 looks for an effect of habitual physical activity identical for both groups. Tests 2–4 were also done for biological age.

Results

The 8 hypotheses, tested by analysis of covariance, did reveal that the explosive arm strength (measured by handgrip) proved to increase significantly ($p \leq 0.01$) in the experimental group relative to the control group (fig. 6).

Series of skill tests were used to evaluate the performance in p-e in a narrower sense at four points during the year. Analysis of these data showed a significantly higher score ($p \leq 0.009$) of the experimental group compared to the control group (fig. 7).

Variables	Test 1	Test 2	Test 3	Test 4
FEV%	n.s.			
Corr. upp. arm diam.	*	n.s.		n.s.
Fat, %	n.s.			
Plate tapping	n.s.			
50-meter shuttle run	n.s.			
Bent arm hang	n.s.			
Handgrip	**	**	n.s.	n.s.
W_{170}	n.s.			

n.s. = Not significant; * $p \leq 0.05$; ** $p \leq 0.01$.

Fig. 6. Results of the analysis of covariance. Tests correspond with figure 5.

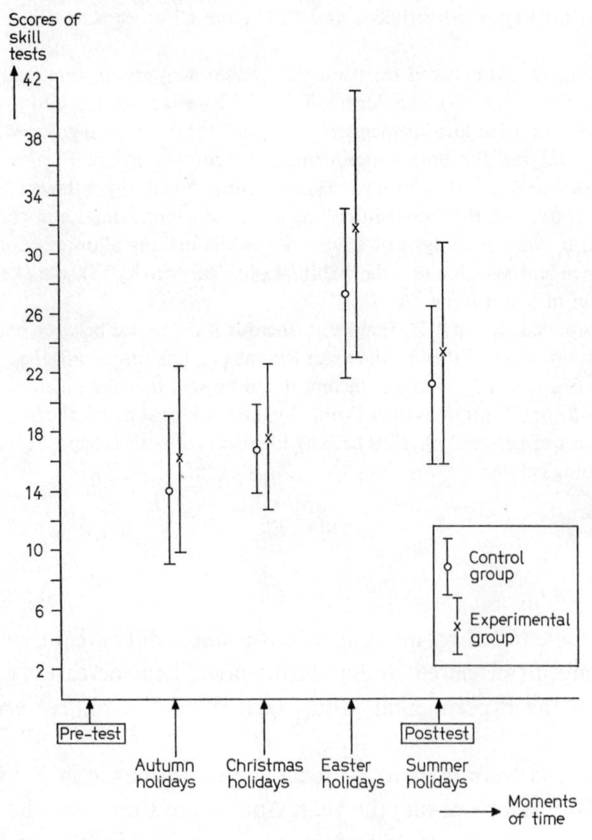

Fig. 7. Mean and standard deviation of scores of skill tests of experimental and control group (ordinate) and periods of measurement (abscissa).

Discussion

Although only the effect of extra lessons in p-e was hypothesized, the influence of habitual physical activity and biological age were also analyzed.

(a) Besides the effect of the extra lessons upon the explosive arm strength (handgrip), an influence of biological age, identical for both groups, could be demonstrated ($p \leq 0.05$). Habitual physical activity had no influence.

(b) Upon the variable corrected upper-arm diameter, a significant effect could be proved in the experimental group ($p \leq 0.05$) caused either by the extra lessons in p-e and/or habitual physical activity and/or biological age. Further analysis revealed that a small, but insignificant ($p < 0.10$) influence of biological age alone is present. Influence of habitual physical activity could not be demonstrated ($p \leq 0.05$). Thus, the difference found in test 1 may be caused by a combined influence of biological age, additional p-e, habitual physical activity and other unknown factors.

Conclusion

In general, it can be stated that the expectations about the effects of two extra lessons of physical education upon 12- and 13-year-old boys could not be confirmed.

In a five versus three lessons a week physical education program, significant increases in p-e skills were found in the experimental group. However, among the other 8 hypotheses, a significant increase in only explosive arm strength could be shown in the experimental group with regard to the control group.

It must be stressed that these results do not have any implication for the evaluation of the regular curriculum of physical education.

References

KEMPER, H.C.G.: The influence of extra lessons in physical education on physical and mental development of 12- and 13-year-old boys. Proc. of a Satellite Symp. of the 25th Int. Congr. of Physiological Sciences, Prague 1971, pp. 212–216 (Universita Karlova, Prague 1973).

KEMPER, H.C.G.; RAS, J.G.A.; SNEL, J.; SPLINTER, P.G.; TAVECCHIO, L.W.C., and VERSCHUUR, R.: Invloed van extra lichamelijke oefening. (The influence of extra physical education.) (De Vrieseborch, Haarlem 1974).

SIMONS, J.; BEUNEN, G.; OSTYN, M.; RENSON, R.; SWALUS, P.; GERVEN, D. VAN, and WILLEMS, E.: Constructie van een motorische testbatterij voor jongens van 12 tot 19 jaar door middel van factor-analysen. Sport *49:* 3–21 (1970).

Voorstel leerplan Rijksscholen voor lichamelijke oefening voor VWO, HAVO en MAVO. (Proposal Curriculum Government Schools for Physical Education.) (Staatsuitgeverij, Den Haag 1968).

WAHLUND, H.: Determination of the physical working capacity. Acta med. scand. suppl. 215 (1948).

WIERINGEN, J.C. VAN; WAFELBAKKER, F.; VERBRUGGE, H.P., and HAAS, J.H. DE: Groeidiagrammen Nederland 1965, NIPG (TNO) (Wolters Noordhoff, Groningen 1968).

Dr. H.C.G. KEMPER, University of Amsterdam, Jan Swammerdam Institute, *Amsterdam* (The Netherlands)

Medicine Sport, vol. 11, pp. 167–172 (Karger, Basel 1978)

Work Capacity, Strength, and Body Measurements of Adolescent Boys in a Special Sports Program Compared to Normal Boys: Initial Comparison

WILLIAM DUQUET and DANIEL GREGOIRE

Laboratory of Human Biometry and Movement Analysis, Vrije Universiteit Brussel, Brussels

In 1972, new curricula were introduced in special schools in Belgium to provide opportunities for in depth instruction in areas such as art, engineering, and sports.

In three schools specially qualified personnel and equipment were provided to conduct the sports program. One of the schools was located in a suburban district of Brussels (Wemmel). The program, applied to 40 boys (table I), consisted of 11 h of physical activity per week, as compared to the previous maximum of 3 h.

A first assessment of physical fitness was made 12 weeks after the start of the program. At the same time, a like assessment was made with a control group of boys of the same age in a school in which not more than 3 h of physical training were given per week.

The test items used in this study have been described in the report of the International Committee on the Standardization of Physical Fitness Tests (ICSPFT) [LARSON, 1974] and in the report on the Belgian Performance and Talent project [HEBBELINCK and BORMS, 1969]. The items and comparisons of the results for the two groups (Wilcoxon sign rank test for significant differences as interpreted at the 5% probability level) are shown in table II.

The absolute values suggested the sports group was shorter in the legs, longer in the trunk, with greater thigh and calf girth and lesser calf fat. However, the tests of significance showed none of these values to be significant. Thus, it was appreciated that the sports and control groups were not radically different in the early phase of the program in most items. The sports group were better performers in three of the four performance tests.

Table I. Course options: orientation guide – option 'sport', 36 h/week

General education, 21 h		Fundamental options				Physical enrichment program	
		scientific courses		language courses		sport	8 h
Non-confessional or religious courses	2	geography	1	second language	3	swimming	1
Physical education	3[1]	biology	1	third language	3	athletics	1
Netherlandish	4	chemistry	1	history	1	gymnastics	1
Second language	3	scientific work	2			games	2
Mathematics	5	third language	2			optional sport	2
Physics	2					theory	1
Sociology and initiation in social economics	2						

[1] All the students in the school are required to take 3 h of physical education classes per week, including the 'sport' option students; however, the latter group are given their lessons separately. Every 14 days, there is also 2 h additional sport activity on Wednesday afternoons for both groups.

In the 1,000-meter run test, no comparison was possible since the sports group ran on a grass course whereas the control group had the advantage of running on a cinder track.

Although there were no significant differences in five of the seven lower body strength tests reported in this paper, the control group performed better in two tests. They showed significantly greater strength in ankle plantar flexion. In the other test (knee extension) the technique differed in that the control group was able to lean slightly backwards on a strength test table, whereas the sports group was restricted to an upright posture by a chair.

Conceivably, it is possible for groups to have identical mean values, yet have mostly different relationships among test variables. In order to examine this possibility, a zero-order product-moment correlation matrix was calculated for each group. The significant correlation values ($p \leq 0.05$) are displayed in table III.

Table II. Summary of initial test scores for experimental and control subjects on selected items and test of differences by Wilcoxon sign rank test

Item	Sports group		Control		Significance
	$\overline{X}$	δ	$\overline{X}$	δ	
Anthropometric					
1 Weight	57.3	8.9	56.9	9.3	0
2 Stature	169.1	8.6	168.8	7.5	0
3 Trochanter height	86.6	4.6	87.5	4.5	0
4 Tibial height	45.1	2.7	45.8	2.5	0
5 Malleolus height	8.1	0.8	8.1	0.8	0
6 Sitting height	87.7	4.9	86.5	3.5	0
7 Calf skinfold	6.3	2.7	8.4	3.8	0
8 Thigh girth	48.5	3.6	48.0	3.7	0
9 Calf girth	34.1	2.7	33.1	5.6	0
Performance					
10 Pulse rec. 10″	41.8	4.0	42.7	4.0	0
11 Pulse rec. 45″	32.8	6.3	36.5	5.6	S
12 Standing broad jump	207.4	19.2	194.0	13.4	S
13 1,000 m	257.1+	28.2	223.9++	38.9	–
14 PWC 170	897.5	178.8	799.3	172.8	S
Strength					
15 Hip flexion	37.7	10.5	39.5	13.6	0
16 Hip extension	40.9	11.2	38.5	9.1	0
17 Leg lift	79.6	22.7	72.2	33.3	0
18 Knee flexion	26.6	8.6	25.2	8.0	0
19 Knee extension	37.1	11.0	49.2	16.9	C
20 Ankle dorsal flexion	23.1	8.1	23.6	9.7	0
21 Ankle plantar flexion	47.1	16.3	54.8	18.9	C

N = 37–40 varying with sporadic absences from school, significance interpreted at 5% probability level with S and C indicating difference in favor of sports or control group and 0 indicating no significant difference.
+ On grass course; ++ on cinder track.

Table III. Significant correlations ($p \leq 0.05$) of variables in control group (lower left) and sports group (upper right)

	1	2	3	4	5	6	7	8	9	10	11	12	13	14	15	16	17	18	19	20	21
1 Weight	–	0.8	0.6	0.6	0.6	0.8		0.8	0.8			0.5			0.4	0.5	0.4	0.3	0.4		
2 Height	0.8	–	0.9	0.8	0.6	0.9		0.5	0.5			0.5		0.4							
3 Troch. ht.	0.7	0.9	–	0.9	0.5	0.6						0.4									
4 Tib. ht.	0.7	0.9	0.9	–	0.5	0.7		0.4	0.4			0.4									
5 Mall. ht.					–	0.7		0.4	0.5												
6 Sitt. ht.	0.7	0.9	0.7	0.7		–		0.6	0.5			0.4		0.5		0.4		0.3		0.3	
7 Calf sk.						0.3	–														
8 Thigh circ.	0.9	0.6	0.5	0.5		0.5		–	0.9			0.4		0.5	0.4	0.5	0.4	0.3	0.6		
9 Calf circ.	0.8	0.6	0.5	0.5		0.6		0.7	–			0.5		0.4		0.4	0.4	0.3	0.4		
10 Rec. 10″										–	0.8					0.3					
11 Rec. 45″										0.8	–		0.3								
12 St. br. jp.	0.4	0.4	0.3			0.5	0.4	0.3				–	0.5	0.5	0.4	0.4	0.5		0.4		
13 1,000 m												0.5	–	0.5				0.4			
14 PWC 170	0.4	0.4				0.4	0.3		0.3			0.4		–			0.3		0.4		
15 Hip fl.	0.5	0.5	0.4	0.4		0.5		0.4	0.5			0.5			–	0.6	0.4	0.5	0.4	0.6	0.5
16 Hip ext.	0.5	0.4	0.4	0.5		0.4		0.4	0.4			0.3	0.3		0.7	–	0.4	0.5		0.4	0.4
17 Leg lift	0.5	0.6	0.5	0.4		0.6	0.3	0.4	0.3			0.6		0.4	0.4		–	0.3	0.3	0.4	
18 Knee fl.	0.3					0.4			0.4						0.5	0.4		–	0.3	0.4	0.6
19 Knee ext.	0.5	0.5	0.4	0.4		0.5		0.3	0.5			0.5		0.4	0.7	0.4	0.4	0.5	–		
20 Ankle d. fl.	0.4			0.3	0.3				0.3						0.4			0.4	0.5	–	0.5
21 Ankle pl. fl.	0.4	0.4	0.4	0.4					0.3						0.5	0.4	0.4		0.5	0.4	–

In general, the correlation coefficients among variables in the sports group were similar in magnitude to those obtained between the same items in the control group. Except for a few values, both correlation matrixes are almost symmetric.

Notable exceptions were for PWC 170 and the 1,000-meter run which were significantly related (–0.487) in the sports group and not significant for the control group (–0.078); and leg lift vs trochanter height (0.482), vs tibial height (0.448) and vs sitting height (0.597) for the control group, whereas the correlations of the same items were not significant in the sports group.

The control group also tended to show higher relationships for hip flexion, hip extension, knee extension, leg lift, and ankle plantar flexion with anthropometric items.

The possibility is suggested that the sports group's training levels may mask the size relationships. However, the magnitude of the correlations was not appreciably higher than $r = 0.305$ which governed significance at the 5% level of probability, and only a trivial common variance is indicated between strength and size items for the control group.

Also, the fact that the sports group is, more than the control group, familiar with and motived for physical performance tests could explain for a part the differences found between the two groups. This theory is supported by the results of the strength tests. No significant differences were found in this less spectacular and for both groups new experience, except for ankle plantar flexion.

Conclusion

In projects such as the Medford Boys' Growth Study [Clarke, 1973] where success in sports often was related with maturity, physique, body size, gross and relative muscular strength, muscular endurance, muscular power, speed and endurance, it was anticipated that the boys in a special sports program would differ from those in a regular school program. This study, however, did not substantiate this assumption. The boys in both programs were similar anthropometrically. Moreover, most of the 210 correlations among variables in the sports group were of a similar magnitude to those obtained between like variables in the control group.

Fortuitously, perhaps, it appears that most of the initial measurements on a sports and a control group are similar enough to be used as initially matched samples in a longitudinal exploration of possible changes associated with a substantially enriched activity program.

References

CLARKE, H.H.: Characteristics of athletes. Phys. Fitness Res. Dig. *3:* 2 (1973).

HEBBELINCK, M. and BORMS, J.: Tests en normenschalen van lichamelijke prestatiegeschiktheid, lengte en gewicht inbegrepen, voor jongens van 6 tot 13 jaar uit het lager onderwijs. (Tests and normscales of physical performance capacity, including height and weight, of boys 6 to 13 years of age from primary schools) (Ministerie van Nederlandse Cultuur, BLOSO, Brussel 1969).

LARSON, L.A. (ed.): Fitness, health and work capacity. International standards for assessment (Macmillan, New York 1974).

W. DUQUET, Vrije Universiteit Brussel, Laboratory of Human Biometry and Movement Analysis, *Brussels* (Belgium)

Subject Index